Social DNA

SOCIAL DNA

Rethinking Our Evolutionary Past

M. Kay Martin

First published in 2019 by
Berghahn Books
www.berghahnbooks.com

© 2019, 2020 M. Kay Martin
First paperback edition published in 2020

All rights reserved. Except for the quotation of short passages for the purposes of criticism and review, no part of this book may be reproduced in any form or by any means, electronic or mechanical, including photocopying, recording, or any information storage and retrieval system now known or to be invented, without written permission of the publisher.

Library of Congress Cataloging-in-Publication Data

A C.I.P. cataloging record is available from the Library of Congress
Library of Congress Cataloging in Publication Control Number: 2018040128

British Library Cataloguing in Publication Data

A catalogue record for this book is available from the British Library

ISBN 978-1-78920-007-2 hardback
ISBN 978-1-78920-757-6 paperback
ISBN 978-1-78920-008-9 ebook

To Eleanor Burke Leacock
1922–1987

Contents

	List of Illustrations	viii
	Preface	ix
Introduction	Some Givens	1
Chapter 1	Perspectives on Anisogamy	23
Chapter 2	First Families	31
Chapter 3	Paleoecology and Emergence of Genus *Homo*	69
Chapter 4	Paleolithic Dinner Pairings: Red or White?	97
Chapter 5	Signature Hominin Traits	115
Chapter 6	Kinship and Paleolithic Legends	157
Chapter 7	Kinship as Social Technology	188
	Epilogue	228
	Endnotes	235
	Bibliography	246
	Index	269

Illustrations

Figures

0.1. Overview of human evolution (copyright John A. J. Gowlett) from Gowlett and Dunbar (2008: 22). With permission of John Wiley and Sons, Inc.   9

1.1. Selfish-gene theory and the origin of female exploitation. By Drew Fagan.   28

3.1. *Homo erectus* lakeshore encampment in the Early Pleistocene. By Drew Fagan.   90

4.1. "Lucius." By Emiliano Troco, oil on canvas, scientific supervisor Davide Persico, private collection.   107

4.2. "Neanderthal Clan." By Emiliano Troco, oil on canvas, scientific supervisor Davide Persico, collection of Museo Paleoantropologico de Po.   107

5.1. Middle Pleistocene *Homo heidelbergensis* butchery site at Boxgrove, West Sussex, England. With permission of Getty Images.   133

5.2. A model of human society based on general systems theory. By Drew Fagan.   154

Tables

2.1. Principal assumptions about *Pan-Homo* social life in androcentric models.   37

2.2. Impact of ecological variables on female feeding strategies and primate social organization. Source data: Wrangham (1979, 1980).   50

 # Preface

I came to the field of anthropology under the mentorship of ethnohistorian Harold Hickerson, content to haunt the card catalogs and cavernous stacks of university libraries for early accounts of preagricultural peoples. My treasure hunts were aimed at uncovering glimpses of aboriginal social organization for hunter-gatherers on three continents and documenting postcontact change in these societies over time. The aggregate picture that emerged for foragers in their most pristine state was one of robust communities, diverse systems of kinship, and a broad spectrum of sociopolitical complexity. For most of these societies, however, their vitality and continuity was short-lived. Genocide, disease, atomism, and cultural dismemberment accompanied the unrelenting advance of European colonialism, leaving them depopulated, displaced, and a shadow of their former selves. Ironically, ethnohistorians, in their efforts to reconstruct the cultures of these peoples, often become the unwitting chroniclers of their sorrows.

During the mid to late 1960s, American ethnologists began to reinvent the concept of cultural evolutionism so roundly rejected by their discipline's founding fathers. Neoevolutionary schemes inevitably commenced with portraits of small, atomistic family bands pursuing a meager living from limited resources in a harsh and unfriendly world. I was struck at the time by the incongruity of memorializing surviving hunter-gatherer societies in these models as living examples of Paleolithic life. Historic foragers, in my mind and experience, were arguably remnant or refugee communities, and unlikely avatars of our ethnographic past. Similarly, the characterization of early human kinship as inevitably male-centered did not square well with what I knew of the ethnohistorical and ethnographic records.

In 1975, I coauthored a book entitled *Female of the Species* with archaeologist and colleague Barbara Voorhies. An overarching premise of this work was that the human evolutionary journey carries the mark of our primate heritage, but that factors in addition to biology play a major role in shaping the gender behaviors and social institutions of both ancient and modern societies. The book presented the results of a cross-cultural study I completed for over five hundred societies that highlighted how ecology and history have influenced the nature and direction of cultural institutions through time.

During this same period, however, the pendulum had begun to swing toward biological explanations for human social origins and primeval kinship. Sociobiological theory placed primary emphasis on genes, and in particular the male genome, as the engine of human evolutionary development. A broadening schism on questions about early human social life developed between biological and social anthropologists, and it persists to the present day. Their at times intractable nature-nurture debates served to not only bifurcate the discipline into dueling camps, but to partition their respective areas of inquiry. Theories on the nature of early human society were largely co-opted by biological anthropologists. In contrast, their social anthropology counterparts, particularly in Great Britain, remained focused on ethnography and synchronic studies, and in subsequent years even moved away from kinship theory.

About this time, my career path detoured from academia to applied anthropology and public administration. It was not until my retirement about a decade ago that I began to revisit the theoretical questions about the nature of human kinship and our social origins that had so long fascinated me. How were male and female reproductive strategies aligned in the first human families? Were ancient hunter-gatherers cast from a single mold, or did these peoples have diverse economic and social forms? Why have societies throughout most of human history organized themselves into groups based on either uterine or agnatic affiliation? One of the interesting revelations upon returning to academic discourse after a somewhat lengthy hiatus was that these big questions had attracted few new answers. It is almost as if one camp had lost interest in asking them and the other camp felt that they had already been satisfactorily addressed.

In pondering possible reasons for this stalemate, I was struck by the way that the field of anthropology had metamorphosed in recent decades and how these changes have impacted communication and intellectual discourse. Those of us trained by second-generation American anthropologists were expected to attain a working knowledge of the four subfields of the discipline. Not uncommonly, one left graduate school with the ability to pass a laboratory practical exam in human osteology, record the spoken word in phonemic transcription, excavate a five-foot square at a prehistoric midden, and find one's way around a kinship chart. One could also attend the annual national meetings of the profession and find concurrent sessions in these four basic subdisciplines; meet and academically engage, face-to-face, with more friends than strangers; and even locate the diminutive Margaret Mead on a crowded mezzanine buried at the proximate end of her notorious shepherd's staff. I believe that this traditional training of anthropologists as generalists in many ways broadened their perspective on the human condition and served to open communication across disparate branches of science.

At the risk of sounding nostalgic, I still have fond memories of lively sit-down discussions and debates with some of anthropology's great generalists on the subjects of evolution, early human society, and kinship systems. Exchanges with Leslie A. White, George Peter Murdock, Elman Service, Marvin Harris, Eleanor "Happy" Leacock, and Ward Goodenough did not always end in agreement, but always left me with a sense of awe about their breadth of knowledge, their openness to ideas, and their perpetual sense of wonder. Their voices, although past, still resonate, and have found their way into this book.

The age of generalist perspectives within the discipline, however, has been substantially diminished. We have moved from a community of generalists to a community of specialists. Whereas in years past professional anthropologists organized themselves into just a handful of subdisciplinary associations and related journals, today's American Anthropological Association lists some fifty specialized sections and interest groups with focused memberships, and over twenty-five separate publications. This proliferation of specialized research foci, while perhaps a natural product of the discipline's growth, also tends to create academic silos that both isolate and insulate their memberships. The trend toward increasing compartmentalization does not bode well for the cross-pollination of ideas. Is anthropology doomed to intellectual myopia and linear thinking?

Two hopeful countertrends provide a potential antidote to long-standing theoretical divides, such as that existing between biological and social anthropologists. First is the wellspring of new cross-disciplinary information that has become available since this debate began. Significant findings are emanating from the fields of primatology, paleontology, geology, paleoecology, genetics, and neuroscience that call for the reconsideration and modification of existing evolutionary theory. It turns out that while some of us were not looking, these parallel bodies of science were addressing many of the same big questions about human evolution, but from different vantage points. These alternative conceptual frameworks are adding key pieces to the puzzle of human origins and human sociality. We need only look. A second positive development is the sea change that has occurred in information technology. Over a relatively short span of time, we have graduated from a universe of typewriters, landlines, mimeographs, paper manuscripts, snail mail, and physical data repositories to personal computers, cell phones, the internet, email, electronic data files, the cloud, e-books, and virtual libraries. Thanks to the digital revolution, scholars of today have literally at their fingertips the ability to not only access an unlimited range of scientific data from their own and other disciplines, but to dialogue with other researchers, some of whom they may never meet, on questions of mutual interest. This is a powerful antidote to linear thinking, and presents both the means and the opportunity to usher in a new generation of generalists.

I have pursued the present work in this spirit. Over the past four years, I have read scores of books and literally hundreds of scholarly journal articles across multiple scientific disciplines. This diverse literature addresses a broad range of questions on hominin evolution and human sociality. The specialized technical data and scientific jargon in many of these sources were admittedly challenging, and some analyses were more difficult to digest than others. But the further along I got on this fascinating journey, the more apparent it became that the nature-nurture controversy in evolutionary theory presents a false dichotomy. New scientific findings inform the social anthropologist that human societies reflect our primate genomic heritage, including not only reproductive behaviors, but how hominid communities have been shaped in similar ways by selective factors in the environment. These findings also inform the biological anthropologist that genes are not the whole story. Their phenotypic expression is influenced by social learning throughout an organism's lifetime. Recent evidence further suggests that phenotypic traits shaped by culture are at least partially heritable and therefore may have played an important role in the pace and direction of human evolution. In short, those who maintain that the human saga has been directed either by the genome or by culture alone are only half right. It is their interface—their complex, synergistic, and dynamic interaction—that accounts for the origins and trajectory of our genus.

Social DNA proposes to peel back the layers of nested hypotheses underlying popular theories on how our species evolved. It examines their interlocking assumptions about male and female natures, mating and intersexual dominance, Pliocene ecology, subsistence and family provisioning, and the extent to which contemporary primates may be taken as avatars of ancient social life. This book systematically disassembles these layers, evaluates the assumptions on which they are based, considers new or competing evidence, and develops alternative hypotheses that lead to novel reconstructions of human sociality.

This analysis is divided into three major parts. The first examines alternative perspectives on the nature of male and female Pliocene apes and how their mating and feeding patterns may have shaped early social groups. New climate and paleontological data are presented that connects early human occupation sites with more humid waterside habitats, greater resource densities, and evidence of dietary breadth, including reliance on aquatic flora and fauna. This reframing of Plio-Pleistocene ecology has profound implications for ancient social life, particularly when such variation in resource distributions is compared with social patterns observed among nonhuman primates. The second part of the book is devoted to a discussion of signature traits shared by members of both ancient and modern hominin lineages that distinguish them from contemporary apes, such as behavioral plasticity, dietary breadth and food sharing, social demography, and cognitive function. *Social DNA* entertains the notion that traditional academic emphases on phylogenetic chronologies, material culture, and gross brain or neocortex size has led to an underestimation of ancient hominin ingenuity, intelligence, and social complexity. A central theme of this discussion is that early hominins were, in some ways, like every other primate, but like *no* other primate. That is, while primates respond in similar ways to similar environmental cues, there are important threshold differences. Contemporary nonhuman species have evolved as specialists, bound by genetic platforms geared to specific niches, whereas hominins evolved as generalists who came equipped with a more diversified gene-culture playbook that broadened their ecological range.

The final section of this book returns to seminal questions about the essence of human kinship. It challenges current theories that tie kinship systems to innate dominance patterns, evolutionary stages, monotypic biograms, or phylogenetic continuity with chimpanzee-like apes. Instead, it proposes that the nature of human kinship systems is not preordained, but rather is the phenotypic expression of epigenetic rules ("social DNA") that optimize the procurement and allocation of fitness-related resources in a given niche. Characteristic patterns of social group formation based on kinship alliance reflect rules for the regulation and distribution of resources critical to survival, such as energy, materials, genes, and information, that are adaptive in specific ecological settings. Matrilineal and patrilineal kinship systems are understood as variable social technologies for niche construction, with distinct architectures for structuring reproduction, labor, and political groups in relation to available resources.

Social DNA is distinguished from other titles on human evolution in its scope of inquiry. New cross-disciplinary research findings are brought to bear on fundamental questions about the nature of Pliocene ecology, Pleistocene subsistence activities, hominin brain evolution, hominin life history changes, and the relationship of inclusive fitness to primeval kinship systems. It is my hope that this book provides a fresh perspective on our biosocial origins and responds to current calls for the creation of an enhanced dialogue across academic and disciplinary boundaries.

I would like to extend special thanks to archaeologist Jon Erlandson, who graciously read and offered comments on parts of this manuscript. The ideas in this book are my own, but have benefitted from his insights on Pleistocene adaptations and the potential role played by aquatic resources in early human societies. Thanks are also due to my anonymous reviewers, who pointed out errors and omissions in the draft manuscript, and whose constructive comments resulted in improvements to this book. I'd also like to acknowledge ethnologist Isabelle Clark-Deces, who provided encouragement at the onset of this effort, and who left us all too soon.

I am grateful to paleoartists Mauricio Anton and Emiliano Troco for agreeing to share their wonderful portraits of early hominins to illustrate this book, and to artist and friend Drew Fagan for his creative genius, technical support, and good humor throughout this project.

Finally, I would like to thank my editors at Berghahn Books, Harry Eagles and Elizabeth Martinez, for their responsiveness and support during the book's review and production.

 Introduction

SOME GIVENS

This book revisits fundamental questions on the biocultural origins of human sociality. What set our ancestors on a separate evolutionary path from that of other apes? What was the nature of primeval kinship and mating? What roles have biology and ecology played in shaping human social groups through time?

Scholarly musings on such questions have populated library shelves for decades. Various theories have been proposed, debated, embraced, rejected, and periodically recycled. In recent years, the scope of inquiry on human origins has been enriched by pioneering research across multiple branches of science. The findings of such studies, however, have not always informed one another in a manner that encourages a re-examination of current evolutionary theory. In other words, establishment of an academic lingua franca that facilitates the creation of integrative models has been elusive.

A primary goal of the present work is to reach beyond traditional schools of thought and foster a cross-disciplinary dialogue on human social evolution. The task is to unravel the fabric of existing theories, explore new independent discoveries on the emergence of our genus, and tie together the myriad threads of this evidence in novel ways. This is arguably an arrogant undertaking, given the impressive lineup of experts who have already offered their insights on the subject. But experts currently disagree, both on the principal drivers of human evolution and on the nature of ancient social forms. Finding new answers to old questions often involves being a contrarian on some levels, and an adventurer on others. It also requires a measure of humility, since all accounts of human social origins are necessarily speculative.

Theories on the origin of society are neither new nor in short supply.[1] It is a subject that has fueled the imagination of both religious and secular philosophers for centuries. It has also spawned a robust body of scientific evidence, primarily in the fields of anthropology, paleontology, and evolutionary biology. Significant fossil and archaeological discoveries have provided the foundation for chronological reconstructions of our biological and cultural journey over the past 5 to 7 million years. Ancient material remains have shed light on the subsistence activities, technologies, settlement patterns, migrations, and cognitive abilities of archaic populations.

Insights into the evolution of human social behaviors, however, are constrained by the natural limitations of what bone and stone artifacts can tell us about the nonmaterial aspects of ancient sociality. How did our ancestors structure mating, labor, food sharing, kinship, and power relationships? To what extent did biology shape these behaviors? How did ecology influence the prevailing social architecture of human groups in both time and space? What role did primate brain evolution and the emergence of symbolic communication play in the trajectory of early social life?

Since we cannot travel back in time to the encampments of our forebears, answers to these questions have to rely on the construction of conceptual models. However, creation of models with clearly defined premises and measurable outcomes is particularly challenging when the task is to explain the origins of phenomena that are no longer directly observable. Interpretations of the fossil and archaeological records have therefore been traditionally combined with observations of contemporary nonhuman primate communities and of historic hunter-gatherers to paint a picture of what Paleolithic social life may have been like.

One of the earliest and most influential of these models was crafted over a half century ago. Commonly referred to as the "hunting hypothesis," it proposes that the first apes to achieve a successful terrestrial existence on two legs established the social mold from which all subsequent forms of humanity were cast. The adaptation of our ancestors to a carnivorous life on the open savanna, it argues, served as a sort of primeval Petri dish for the germination of a distinct complex of traits—traits that predisposed early humans toward a uniform pattern of reproductive behavior, labor division, and kinship organization. Early sociobiological models argued further that this trait complex was so intimately connected to survival that it became genetically imprinted. In this view, the social life of Paleolithic peoples conformed to a standard template, one dictated by a biogram that not only was perfected by natural selection in ancient times, but continues to dominate the reproductive and social patterns of modern humans.

This monotypic model of human social evolution has enjoyed such popularity over the years that it is sometimes referred to as "the standard narrative." The image of early humans organizing themselves into small male-centered family groups and wandering over parched landscapes in pursuit of sparse game is remarkable by its persistence, particularly in light of subsequent research. In recent decades, fossil, archaeological, and ethnographic records have expanded exponentially, along with our understanding of primate sociality. The hunting hypothesis and its implications for human social life have come under numerous challenges, but such debates have taken place largely within the confines of specific disciplines, such as anthropology and paleontology. Meanwhile, several ancillary branches of science, such as climatology, paleoecology, paleogenetics, and neurobiology, have been exploring their own avenues on the conditions affecting the evolution of our genus, many of which have significant implications for the nature of ancient social life. These new research findings are impacting traditional notions about what early humans ate, where they lived, how their brains evolved, and how ecology and the reproductive strategies of the sexes may have impacted the nature of social groups. In short, many of the fundamental assumptions of existing models are beginning to erode, but robust cross-disciplinary dialogue on these issues has lagged.

Rethinking human social origins is an exercise in collaborative inquiry. Such is the challenge of this book. Recalibrating current evolutionary paradigms is difficult, in part because it often requires a departure from academic comfort zones. Institutions of higher learning create disciplinary and subdisciplinary silos—each with their own legacies of specialized knowledge, jargons, and world views—that constrain the cross-fertilization of ideas. Experts don't always talk to one another or, worse, become vested in their own viewpoints and stewardship of specific schools of thought. A major historical divide, for example, has existed between biological and social anthropologists with regard to the origin and nature of human kinship systems. If productive dialogue is sometimes constrained by banter and debate within individual disciplines, communication problems are compounded by the isolation created by institutional boundaries between them. The need to establish a more comprehensive dialogue on human social origins has recently been highlighted by Callan (2008) and others, such as Mills and Huber (2005), who have proposed the concept of intellectual "trading zones" to foster the communication of ideas across traditional academic disciplines.[2] In short, progress on theoretical questions such as the structure of ancient social life requires a lowering of technical and research boundaries and a more effective way to disseminate and integrate relevant data among scholars from widely disparate fields.

Progress on the refinement of conceptual models also requires a reassessment of cultural and individual biases. Scientific inquiry is an imperfect exercise. It assumes that the scholar approaches the examination of a problem dispassionately, developing insightful hypotheses and then objectively unraveling certain truths through a process of vigorous inquiry or testing. While some disciplines, such as mathematics, naturally lend themselves to the discovery of empirical proofs, others struggle to assemble fragmentary bits of information into some kind of formula or model that purports to explain extant conditions or end states. That assembly process often draws on an assortment of facts, hunches, and a priori biases, the segregation of which may be murky for both the scholar and the intended audience.

Ideally, authors of theoretical books such as this one should be required to devote their first chapter to a declaration of their underlying assumptions and predilections. This exercise would facilitate the author's own awareness of the preconceived notions and agendas they bring to the table. It would also key the reader to factors that are likely to color the author's focus of inquiry, selection of data, and conclusions drawn. An ancillary benefit of this early-warning system, of course, is that it would also provide an opportunity for the reader who disagrees with the book's initial premises to return it to their retailer unread for a full refund. In most cases, however, it is a fair bet that readers would probably welcome a clear exposition of an author's starting point and the opportunities for constructive debate that such honest dialogue provides. This introduction is written in this spirit.

The book thus begins with a summary of assumptions on eight general topics that have influenced this writer's approach to the evolution of human sociality. This initial discussion draws attention to key issue areas in which recent cross-disciplinary research is both augmenting and redirecting our understanding of Paleolithic social life. Each will be discussed briefly, and an effort made to the elucidate how these baseline concepts are reflected in subsequent chapters.

Genes, Epigenesis, and Social DNA

The fundamental assumption of the current work is that modern humans (*Homo sapiens sapiens*) are the product of gene-culture co-evolution spanning at least the past 5 to 7 million years. Current knowledge about how the evolutionary process works has been advanced by three major milestones. The first was publication of Darwin's *The Origin of Species* (1859) and *The Descent of Man* (1871). His revolutionary concepts moved questions about human origins from the realm of philosophy and myth to the discipline of science and established natural selection as the cornerstone of evolutionary biology.

The second milestone was development of the modern science of genetics. Genes were identified early in the twentieth century as the units of heritable traits, and seminal works, such as Dobzhansky's *Genetics and the Origin of Species* (1937), laid the foundation for understanding gene flow through time and space. Discovery of DNA structure in 1953 revealed the molecular mechanics of how traits are transmitted. Later advancements in DNA sequencing in the 1970s and the 2003 reconstruction of the human genome are now allowing us to probe relationships among the ancient lineages of our family tree more deeply, often with surprising results.

The third source of enlightenment on evolutionary processes was the emergence of the field of sociobiology, officially launched in 1975 by E.O. Wilson's *Sociobiology: The New Synthesis*, and its sequel, *On Human Nature*, in 1978. These works helped to establish an interdisciplinary approach to understanding the evolution of heritable physical and social traits in all animal species, including humans. The sociobiological movement overcame initial criticisms of biological reductionism and genetic determinism and went on to spur a wealth of new research that continues to flourish decades later.

The field of sociobiology hosted lively internal debates as well, not the least of which concerned the locus of natural selective processes. One school of thought places primary emphasis on the theory of kin selection. This concept, which originated in the earlier works of biologists Hamilton (1963, 1964), Trivers (1971, 1972), and Alexander (1974), proposes that individual organisms maximize their own reproductive success or "inclusive fitness" by behaving altruistically toward close kin, weighted by the degree of genetic relatedness. The theory, also known as "Hamilton's Rule," was supported by a mathematical formula that calculated that altruism will develop to the extent that the benefit to the recipient times the degree of kinship to the altruist is greater than its cost.

Kin selection as the principal driver of human social evolution gained widespread acceptance among biologists, including E.O. Wilson, for about four decades. Commencing in 2010, however, a series of coauthored papers by Wilson and others challenged the mathematical and biological validity of kin selection theory as an explanation for the evolution of advanced social behavior.[3] In its place has emerged the concept of multilevel selection, in which the evolutionary dynamic is seen as operating simultaneously at both the individual and the group levels. As proposed by Wilson (2012: 162), individual selection is based on competition and cooperation *within* groups, and promotes selfish behavior by its members, whereas group selection is based on competition and cooperation *between* groups, which promotes internal altruism. Wilson views human evolution as a product of these conflicting selective processes in which the interests of the individual must be balanced against the interests of the larger collective.

Multilevel evolutionary theories assume that groups that develop internal structures for cooperative endeavors have adaptive advantages that accrue to their membership. Robin Dunbar (2008), for example, proposes that individuals enter into social contracts to enhance their prospects for survival and reproductive success. He goes on to caution, however, that multilevel selection should not be confused with "group selection:"

> In kin selection, the final arbiter of what happens is the gene, not the group as an entity, and hence it requires no new mechanism of evolution other than standard Darwinian processes. . . . In multilevel selection again, the unit of evolutionary cost-accounting is the gene, and not the group. Group-level processes are intended to facilitate the successful replication of the individual member's genes, not to facilitate the successful replication of the group. The distinction is subtle, but important. (Dunbar 2008: 147)

Richard Dawkins, in *The Selfish Gene* (1976) and *The Extended Phenotype* (1982), cast the gene as the sole protagonist in the evolutionary drama, discounting the role of both individual organisms and groups in the natural selection process. He proposed that genes and their respective alleles act in their own self-interest, programming the organisms in which they reside to behave in a manner that optimizes their frequencies in the gene pool. Genes effectively hitch a ride on human "survival machines," moving their hosts in directions that foster their own replication. In this view, adaptations represent the phenotypic effects of genes to reproduce themselves in future generations. All of this is seen as occurring beyond the conscious recognition of individuals, who are essentially temporary vehicles for gene replication. Dawkins also accounts for the role played by culture in human evolution with the parallel concept of "memes," which are proposed as the units of cultural inheritance. Memes are crafted on the same genetic metaphor, competing with others in a meme pool. Like genes, memes have phenotypic effects, and are thought to be naturally selected by virtue of their successful replication.[4]

The significance of gene-centered theory for models on human origins is twofold. First, it proposes that sociality is (unconsciously) pursued by individuals largely on the basis of self-interest. Degrees of genetic relatedness become the floating calculus for cooperation and competition among individuals, who assemble and participate with others in a tit-for-tat world. Society thus defined becomes a collection of vying gene carriers—a procession of self-serving males and females, kin and non-kin, marching to the zero-sum drum of genomic replication. Second, some applications of gene-centered theory assume that characteristic reproductive strategies and associated phenotypic behaviors, such as dominance, aggression, or parasitism, have become imprinted into our DNA as a kind of species-specific biogram. In other words, ancient and modern humans, in their quest for self-replication, have been pre-programmed to favor certain behaviors and types of social organization to the exclusion of others.

While recognizing that the inclusive fitness of individuals rests on the replication of their genes, the present book will argue that the reproductive success of ancestral humans was not only enhanced by, but *reliant on* their ability to forge cooperative relationships and function effectively within social groups—communities that typically extended beyond the circle of immediate kin to include the broader membership of a breeding population. Humans are not solitary breeders, but group-bonded primates. Ancient human social groups were more than just a collection of individuals with whom to play out one's genetic hand. The alliances and cooperative relationships on which they were based provided an internal division of labor for the acquisition and distribution of fitness-related resources that enhanced the reproductive success of all group members—a characteristic referred to by Wilson (2012: 133) as *eusociality*.

The process of evolution has been understood as involving the interaction of natural selection and genes that are either inherited through DNA or arise via random mutations. However, the recent discovery that an organism's phenotype may be modified by a myriad of nongenetic factors, and that such phenotypic variants are themselves heritable, is transforming the field of evolutionary biology. The process by which this occurs, *epigenesis*, modifies the expression of genes without changing the underlying molecular structure of DNA. A new branch of theory, referred to as the *extended evolutionary synthesis* (EES), proposes that heredity is a developmental process influenced not only by genes, but by an organism's cumulative interaction with its chemical, natural, and social environments. Epigenesis provides a source of nonrandom phenotypic variation once thought reserved for random mutations. Animal experiments have also demonstrated that epigenetic inheritance allows for the storage and transmission of learned information and provides the flexibility for organisms to modify their phenotype in response to rapid environmental change. EES proponents maintain that an organism's niche construction (its selection and modification of its habitat and environmental resources) also affects the direction of evolution by modifying natural selective factors. In other words, the evolutionary process is more complex than simple genomic theories propose.

This perspective on the critical role played by epigenetic traits will find expression in the chapters that follow. What separated early humans from other apes was their gradual emancipation from purely hardwired responses to reproductive and subsistence challenges through a combination of epigenesis, behavioral plasticity, and cortical expansion. Instead of slavishly following an innate prescription or biogram for sociality, epigenesis provided a "soft inheritance system" that allowed humans to alter their behaviors and the structure of their social groups in response to stochastic environmental conditions. Wilson's concept of "epigenetic rules" (2012: 193) parallels what is referred to here as *social DNA*. Social DNA consists of the underlying rules for characteristic human behaviors and social forms that have been selected and replicated over time by virtue of their role in enhancing reproductive success. While they provide a general framework for the human experience, the phenotypic expression of these rules is not unitary or preordained, but is sufficiently plastic to respond to external change.

One of the challenges in unraveling current conceptual models of human social origins is their tendency to meld ideas on evolutionary prime movers, reproductive strategies, sexual dominance, altruism, and social forms into a hardwired genomic platform. An effort is made in the succeeding chapters to deliberately separate these issues for closer examination and discussion.

Chronologies, Crania, and Traditions

Eugene Dubois's unearthing of a million-year-old *Pithecanthropus erectus* skullcap in Java in 1891 inspired generations of paleontologists and amateur rock hounds alike to find the "missing link" connecting apes and humans. A century and a half later the growing fossil record has enabled a general reconstruction of human evolution (Figure 0.1). Fossil remains have typically been grouped into evolutionary chronologies based on their provenance, and on characteristics such as their skeletal and cranial morphology, dentition, estimated brain size, and associated stone tool traditions.

While resultant phylogenetic trees vary somewhat from one another, most scholars propose that hominins[5] evolved in Africa between 5 and 7 million years ago (ma) from among a heterogeneous stock of

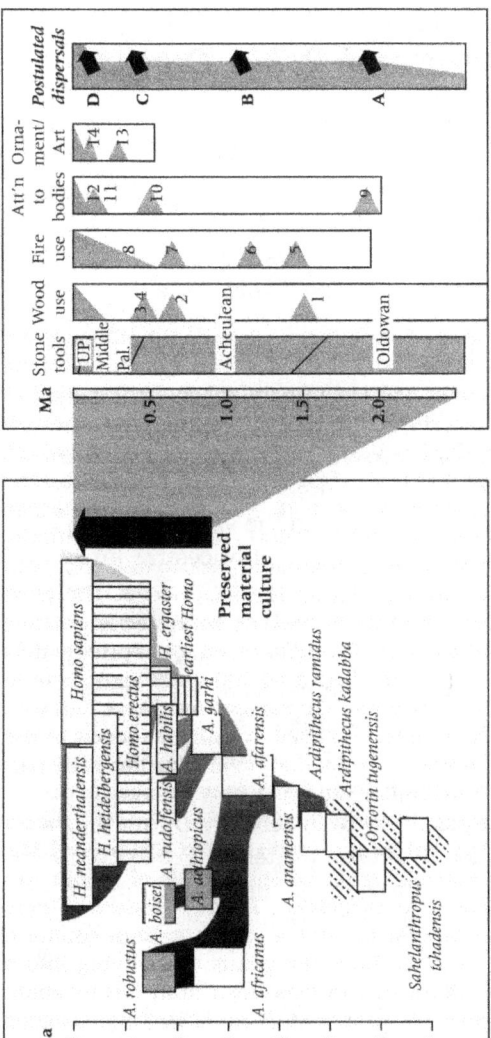

Figure 0.1. Overview of human evolution (copyright John A. J. Gowlett) from Gowlett and Dunbar (2008: 22). With permission of John Wiley and Sons, Inc.

bipedal Pliocene apes. Potential candidates for the earliest primates on the human family tree include a diverse clade of tool-using apes known collectively as australopithecines, as well as more ancient and less robust forms, such as *Ardipithecus ramidus*. Members of the genus *Homo*, distinguished by their larger bodies and brains and by their fully omnivorous diet, are generally recognized as emerging around 2 ma. These first humans are exemplified by fossils such as *Homo habilis* and by multiple waves of *Homo erectus* populations that radiated throughout Eurasia over the succeeding millennia. A prevailing view is that advanced forms of *H. erectus* diverged around 400–500 thousand years ago (ka) into separate lines in Europe (*H. heidelbergensis*) and Africa (*H. rhodesiensis*). These populations ultimately led to the parallel emergence of Neanderthals (*Homo sapiens neanderthalensis*) and anatomically modern humans (*Homo sapiens sapiens*), respectively, by at least 200 ka.[6] This progression of hominin types is associated with evolutionary milestones such as refinement of the infracranial skeleton, increasing encephalization, dietary reliance on animal flesh, the invention of fire, and increasingly complex stone tool technologies.

Such evolutionary reconstructions provide the essential backdrop for current theories on human social origins. Recovered fossil and cultural remains have been utilized as windows on the subsistence activities, cognitive abilities, and social organization of ancestral humans through time. A cautionary note, however, is that our perspective on Pleistocene lifeways is based on material records that are still quite fragmentary, and that are subject to revision with the next great find. Recent discoveries suggest that ancient hominins were much more diverse than previously realized. Fossil specimens do not always fit into the tidy boxes of existing chronologies. In short, when looking back, it is important to remain open to new information and new interpretations—to know how much we don't know. There are points in this book where the reader will be asked to entertain alternatives to popular theory where conclusions have been drawn based on only limited data or, alternatively, where new information compels us to perceive ancient social life in different ways.

For example, modern humans are ensconced at the pinnacle of evolution—as the brightest and the best—while Neandertaloid, *Homo erectus,* and other archaic hominins have often been portrayed as brutish, dim-witted, and doomed to extinction at the hands of more intelligent or technologically advanced peoples. As will be argued later in this book, although hominin brains got bigger through time, so did hominin bodies. Relative brain size is being increasingly questioned as a reliable indicator of ancestral hominin intelligence. Recent neurobiological research suggests that the key to cognitive abilities may lie, instead, in the neuronal density, circuitry, and conduction velocity of gray matter, factors that are not discernible through external cranial measurements or endocasts of fossil specimens.

Similarly, the chronological or cognitive pedigree of fossil hominins cannot be reliably equated with or pigeonholed by the type of lithic traditions with which they made their living. Simple pebble tools and flakes, for example, were utilized for millions of years alongside or in the absence of stone tools fashioned by more refined knapping techniques. A rule of thumb is perhaps that ancient peoples used tool types that "worked" in their ecological niche, regardless of their antiquity or point of origin. Moreover, nonlithic traditions based in perishable organic materials such as wood and fiber are underrepresented in the archaeological record, but may have provided the basis for alternative ecological adaptations. Thus, it is less useful to characterize ancient populations by their tool types—as Oldowan, Acheulean, or Mousterian "peoples"—than it is to examine the range of adaptations potentially associated with these and other material technologies.

The tendency for chronologies to put ancient hominins into discrete boxes has a long tradition in paleontology, where each new fossil discovery is frequently celebrated with a unique phylogenetic designation as a distinct species. Historically, such specimens have been grouped into evolutionary stages that connote a linear progression of increasing complexity over time. Passage from one stage to the next is often perceived through the lens of replacement, with lesser species absorbed or overrun by more advanced species through mass migration, technoeconomic prowess, or superior intelligence.

An alternative theory proposed early on by Ernst Mayr (1950), and more recently adopted by Wolpoff (1989) and Finlayson (2014), is that there was never more than one species of *Homo* at any one time in our evolutionary past.[7] In other words, once the threshold to the genus had been crossed, subsequent evolutionary changes were largely a matter of degree rather than kind. In this model, *Homo sapiens erectus* represents a single polytypic species that established the gene-culture foundation for all subsequent members of the genus, occupying the entire breadth of temperate latitudes in Africa and Eurasia (a region Finlayson refers to as the "Middle Earth") for about 1.5 million years. The single-species theory recognizes that separation of hominin clades by geographic barriers and by climate change events led to the periodic isolation of gene pools, the proliferation of distinct lineages, and periodic extinctions. It differs from more traditional models, however, by arguing that such separations were insufficient to produce speciation. In other words, significant gene flow occurred throughout the Pleistocene, allowing members of diverse *Homo* lineages to periodically reconnect, mate, and produce fertile offspring.

A major criticism of the one-species theory lies in the degree of diversity observed in the accumulating hominin fossil record. A wide variety of hominins appear to have lived contemporaneously throughout the Pleistocene, alternating periods of migration and inbreeding with periods of relative isolation and genetic bottlenecks. Paleontologists disagree on where to draw species boundaries among disparate lineages of ancient hominins, but are increasingly reticent to view the course of human evolution as a linear progression of forms—as a single main line flanked by a number of side branches or dead ends. Rather, our evolutionary pathway may more closely approximate a braided stream, the various branches of which periodically diverged, crossed over, and reconnected throughout the Pleistocene.

Boaz and Ciochon (2004: 166) have suggested that it may be more appropriate to replace the concept of gene pool with *gene sea,* across which genes flowed subject to the currents, waves, and eddies created by climate change and natural selection. Adaptive changes have taken place gradually among closely related populations, creating clinal gradients over time and space. *Homo sapiens* is now, and has always been, a polytypic species. The recent sequencing of Neanderthal and modern human DNA lends credence to the notion of ancient gene exchange and of population replacement through hybridization. Resources and opportunities permitting, Pleistocene hominins made love, not war. The braided stream model of gene flow and speciation will be favored in this book.

Water Drives Ecology

The course of human evolution is inextricably linked to water. Daily access to adequate drinking water is a biological imperative and the primary limiting factor that has shaped how and where ancient peoples lived and prospered. Discussion among paleoanthropologists has been focused in two principal areas: (1) water as a critical element of the early hominin habitat that accompanied arboreal abandonment and expansion into the savannas; and (2) the impact that climatic events and associated changes in water distribution and availability had on the demography and lifeways of Pleistocene populations.

The literature abounds with theories on the abandonment of arboreal life. There is general agreement, however, that the first hominins to venture from the warm, moist environment of the forest canopy did so on the margins of transitional habitats that were defined by water sources. Finlayson (2014: 22–24) surveyed the habitat elements common to known fossil sites in order to better characterize what he termed the essentials of early hominin real estate. The vast majority were found to be associated with settings that combined "shallow water, trees, and open (treeless) spaces." Terrestrialism expanded the range and types of potential food sources for initiates, gradually converting them from tacit fruitarians into omnivores. This bipedal vanguard, commonly associated with a wide variety of Pliocene apes, is assumed in popular theories to have been followed by more advanced proto-humans that ventured farther into the surrounding countryside to procure sources of animal flesh—an accomplishment often credited with leapfrogging early hominins into the genus *Homo*.

Water, or the absence thereof, is seen by Boaz (1997) and Finlayson (2014) as a driver of increasing sophistication in early humans. Pleistocene climatic swings were accompanied by dramatic ecological changes, including the retreat of tropical forests, increasing aridity, and desertification. Such conditions are proposed to have selected for complex reasoning and strategies necessary for survival in the face of diminishing water supplies. Prevailing theories about life in the Pleistocene argue that early hominins, tethered as they were to sources of water, were forced to increase their mobility and geographical range, limit their group size, adopt new technologies, and/or embark on migrations to more favorable habitats for sustenance.

Such theories acknowledge the critical importance of water sources for quenching hominin thirst. But notably, they are generally silent about the extent to which rivers, lakes, marshes, and marine shorelines also contained the necessary food sources to fill their bellies. Such well-watered habitats, which support a wide variety of flora and lipid-rich fauna, were variably distributed throughout Africa and Eurasia even during the waxing and waning of glacial events. These premium habitats were theoretically capable of supporting sizeable populations on a seasonal or year-round basis. Not coincidentally, this is precisely where hominin fossils are most frequently found. This book will examine recent paleoecological research and archaeological evidence that underscores the importance of mosaic habitats and aquatic resources in early human adaptations.

The implications of well-watered habitats and resource diversity for early human social life are profound. Periods of extreme aridity did occur during the Pleistocene. Lakes and rivers in some regions did dry up. But surviving populations were necessarily opportunistic. It is equally probable that, as an alternative to settling for a Spartan existence on the arid savanna, enterprising hominins simply followed existing waterways and coastal highways to more favorable habitats, or shadowed the advance and retreat of glacial ice seeking the "ecological release" of untapped environments. This book will re-examine what we know and what we think we know about the hominin Plio-Pleistocene diet and how these assumptions color our vision of early human communities.

Ecology Drives Social Forms

Traditional theories on human social evolution often weave the mating and economic dimensions of reproductive success into a single fabric, one that assigns a primary role to male inclusive fitness. In this view, ancient genetic propensities that wed male sexual and economic dominance are proposed to define not only the dawn of human kinship and the division of labor, but much of what drives contemporary reproductive behaviors. The resultant social forms proposed as primal markers of humanity thus become monotypic or one-dimensional, with exceptions viewed as aberrant or unnatural. Such models create a single Paleolithic prototype that is largely impervious to ecological variation.

In contrast, the multilevel selection perspective advanced in this book provides a framework for understanding the factors that impact the inclusive fitness of both sexes and the variable structuring of social relationships based on common kinship. A basic assumption is that human sociality evolved as a vehicle for reproductive success. Selective factors operated to increase fitness by not only structuring mating behaviors and relationships among close kin, but by structuring relationships among community members in a manner that optimized their procurement and distribution of fitness-related resources, such as energy, materials, genes, and information. Hominins evolved epigenetic playbooks, or social DNA, the phenotypic expression of which calibrated social systems with the nature and availability of critical resources in a given niche. The architectural types of human social groups have been limited in time and space and have conformed to a finite set of rules. Mating relationships may be predominantly monogamous, polygynous, polyandrous, or polygynandrous.[8] Similarly, social forms that define economic units and access to resources are of limited types. For most of prehistory, they have been based on the recognition of either uterine (related female) or agnatic (related male) kinship. While the number of social DNA variants ("epialleles") is small, their phenotypic expression is plastic and importantly linked to ecological conditions.

Pleistocene habitats on the African and Eurasian continents were both dynamic and diverse. Some were arid, some well-watered. Some offered year-round abundance, others only seasonal or scant resources. All were subject to climatic events that choreographed the changing demographics of species. This book proposes that hominin populations occupied a wide range of niches throughout their evolutionary development. Social DNA or epigenetic mechanisms of inheritance provided evolving humans with the capacity to flex their social strategies in characteristic ways to meet the challenges presented by changing landscapes. The social forms that structured these adaptations, namely, multigenerational kinship units, are of limited types. Uterine and agnatic social organization have distinct demographic, social, and political consequences that correlate with ecological factors. Their phenotypic expression should therefore be predictable in general outline. The final chapter of this book explores the rise and fall of Paleolithic kinship groups and the ecological dynamics on which they may have been predicated.

False Prototypes

If *Homo sapiens sapiens* were the only primate to survive to modern times, the task of breathing life into ancient fossil remains would be doubly challenging. Our ability to observe the behaviors and social life of contemporary apes and monkeys provides an opportunity to identify traits that may be markers of our common ancestry, and hence perhaps equally shared with proto-human populations. Earliest attempts to utilize observations of primates in the wild as a window to societal origins were commonly based on savanna baboons, not only because they were the most highly studied, but because they were thought to occupy human primordial habitats. Lionel Tiger (1969), for example, envisioned baboon male dominance hierarchies as a genetically based precursor to male bonding requirements of the hunt, activities claimed as pivotal to human evolutionary development on the open savanna.

In the succeeding decades, additional studies have more clearly defined the complexity of baboon social organization, along with that of chimpanzees, bonobos, and a variety of Old and New World monkeys. Such studies have provided new information on primate sexuality, male and female dominance hierarchies, and social networks, along with an appreciation of how such behaviors respond to ecological factors. While patterns common to contemporary nonhuman primates provide valuable insights for reconstructing our ancient past, there are some cautionary notes.

Models based on a single species observed in a single habitat are subject to the whims of the author's choice, and therefore may lead to false generalizations or conclusions. The same may be said about reliance on modern apes as avatars of our ancient past. It is important to remember that all contemporary nonhuman primates have been forced into marginal habitats that are not representative of Plio-Pleistocene ecological conditions. Some apes, such as gorillas and chimpanzees, have been refugees for millions of years, retreating and adapting to the isolation of shrinking African tropical forests. These and other primate species have also suffered repeated encroachments by *Homo sapiens*, conditions that may enhance competition for resources and fundamentally alter their behaviors and the structure of their communities.

The same cautionary note applies to models of Paleolithic society, which have historically been based on contemporary hunter-gatherers. Virtually every anthropology text on human cultural evolution portrays the !Kung Bushmen, Hadza, or Australian aborigines as examples of what preagricultural life was like in the Pleistocene. The predominant theory has been that the earliest humans emerged with a dietary reliance on animal flesh and ranged into the open, semi-arid savanna as small, highly mobile hunting bands. The so-called hunting hypothesis grew out of conferences and symposia in the 1960s and its standard-bearer, the patrilocal band, was long memorialized in the anthropological literature as the earliest stage in human cultural evolution.[9]

One of the cautions of viewing lifeways of contemporary hunter-gatherers as windows to our Paleolithic past is that they are cultures in crisis. Khoisan-speaking groups such as the Bushmen were once dominant in eastern and southern Africa, but have been sequentially displaced from favorable ecological niches for centuries by Nilotic herdsmen, Bantu-speaking horticulturalists, colonial Europeans, and contemporary Africans. Similarly, various aboriginal ethnic groups in Australia were not only displaced from resource-rich coastal areas by Europeans, but decimated by genocide and disease. In both cases, depopulation and forced retreat into increasingly marginal habitats has systematically dismantled indigenous social and economic structures. While such societies can serve as useful models for how humans survive conditions of population loss and resource scarcity, the niches they currently occupy and to which they have been forced to adapt cannot be assumed to have dominated the Pleistocene landscape. These populations, like many nonhuman primate groups, are also refugees. The ethnographic present does not represent the ethnographic past.

A related critique of such prototypes is that the diversity of both contemporary and ancient hunter-gatherer adaptations has been systematically ignored (Martin 1974). Why, might one ask, when searching for historic or prehistoric hunter-gatherer examples that may potentially shed light on how Pleistocene populations lived, are the inhabitants of more resource-rich environments not considered? While climate change and increasing aridity were certainly factors affecting the evolution of our species, the impacts of these changes were unevenly distributed in time and space. Some areas of Africa and Eurasia may indeed have experienced rhythms of aridity that mimicked the challenges of today's Kalahari and Australian deserts. But there were also vast areas of the Old World that remained well-watered and optimal for occupation by humans and other animals. Opportunistic members of all species followed the geography of these rhythms to their advantage. Those that did not perished. It is interesting to contemplate that if our ancestors were among those who chose to tough it out on the arid plain, our lineage may have ended up as just another in a long line of extinctions. The present book will entertain the notion that the ecological niches early hominins occupied and the adaptive strategies they employed through time were not uniform, but diverse. The Ice Age was a challenge that selected for ingenuity and plasticity, and is what made us who we are.

Aggress or Coalesce?

Behind every theory on hominin social origins is an underlying set of biases on what constitutes basic human nature.[10] What are the innate propensities that govern intra- and intersexual behaviors, reproductive strategies, social structure, and intergroup relations? As noted earlier, some inclusive fitness models see society as a collection of individuals who engage in cooperative and competitive actions solely in pursuit of their own ends. This portrait of human reproductive behavior congers images of situational loyalties, dominance, subordination, parasitism, deceit, and betrayal. Other origins theories attribute the darker side of human nature to adaptations thought to have accompanied the shift of early hominins from arboreal to terrestrial life. For example, some argue that hunting on the open savanna turned our ancestors into bloodthirsty killer-apes, fierce defenders of territories, and combatants in endemic warfare over mates and scarce resources. A penchant for aggression and violence has also been proposed as an enduring human trait by pointing to these behaviors in contemporary chimpanzees.[11]

At the other end of the spectrum are theories that suggest that cooperation had greater currency in the evolution of human sociality than competition or aggression. Namely, what separated our ancestors from those of contemporary apes was their liberation from primitive limbic system responses to external stimuli and the attendant modulation of hormonally driven behaviors. Current evidence suggests that cortical expansion and a reorganization of brain function was selected for among early hominins, along with the corollary evolution of behavioral plasticity. The increasing ability to interpret stimuli in light of past experience and to apply reasoned, nuanced responses provided hominins with the necessary social tools for empathic and cooperative behaviors. In short, the success of our ancestral lineages relied on the ability to create win-win rather than win-lose scenarios. As recently noted by Clancy (2017), the emergence of our genus may be more accurately described as "survival of the friendliest." Survival and reproductive success came to those who learned how to get along and to modify their strategic alliances to meet the challenges of changing environmental conditions. This perspective will be favored here.

While behavioral plasticity greatly expanded the social repertoire of early hominins, it did so without sacrificing the option for outlier responses. Humans are capable of both extreme empathy and extreme violence. Where threats to the welfare of offspring or access to critical resources present themselves, limbic system behaviors may rise to the occasion. Such responses, however, should not be regarded as the essence of human nature, nor as a rationalization for the inevitability of human aggression, warfare, or systems of inequality. What *is* innate is our ability to gauge responses appropriate to stochastic events. If negative or defensive behaviors have become more frequent in the Holocene and Anthropocene epochs, so too, perhaps, have the environmental conditions that trigger this ancient survival response.

Science and Storytelling

Theories on the nature of human nature and societal origins have always been an odd mix of empirical data and fanciful storytelling. I still have vivid memories of a lecture I attended in the 1960s in which my anthropology professor took a stumbling, bent-knee stroll across the stage to demonstrate the typical locomotion of ancient man. The fact that the speaker's cranial morphology had strikingly Neandertaloid characteristics only served to add to the perceived drama and authenticity of the performance. Subsequent discoveries of additional, nonarthritic Neanderthal skeletons, of course, quickly put this misconception to bed. But not all notions on what life was like on the long road to humanity are so easily tested and modified. In fact, most theories are forced to address many open questions about the course of human evolution, the answers to which will probably remain unknowable.

Why is this? The tangible evidence we have in hand to trace our evolutionary pathway is fragmentary, and may remain so. A tooth here, a tibia there, and if we're lucky, a skullcap or a complete infracranial skeleton. Exceptional finds, such as the multiple skeletons recently unearthed from the Dinaledi Chamber, South Africa, while enlightening, often pose more questions than answers.[12] And so it goes with other sources of data. Geology and paleoecology tell us something about the earth's past climate and physical environment, but also highlight the prospect that many ancient occupation sites may have been forever lost to rising sea levels. Stone tools give hints about how early humans made a living, but other components of ancient toolkits that could provide major insights, such as items made of wood, bone, and fiber, have long since turned to dust. Similarly, contemporary primates provide glimpses of ourselves and perhaps of a common ancestor, but differ from our proto-human forebears in ways we can only imagine.

And imagine we do. Evolutionary biologists and paleoanthropologists struggle to assemble a complex human-origins puzzle that has unknown dimensions and many missing pieces. Consequently, scientists use their imagination and regularly make up stories (aka models) that paint a more complete picture from the fragments they possess. Multiple stories can be generated from the same evidence or set of facts, and the relative veracity of their competing plots is vigorously debated. A select few, however, are inevitably elevated to the status of academic dogma. These stories are told and retold for decades, and may gain traction for reasons beyond the evidence presented, such as the author's reputation or the extent to which their conclusions correspond to prevailing cultural stereotypes or popular views of what constitutes "common sense." But, ideally, dogma is eventually challenged and replaced with new ideas and syntheses that take us closer to an understanding of our roots.

One of the primary challenges facing human-origins storytellers has been the intrepid duo of ethnocentrism and "anthropodenial."[13] Nothing offends scientists more than the suggestion that they approach their subject with less than an open mind. But scientists are human, and humans are creatures of their own culture. When it comes to visualizing what ancient peoples were like, theorists are notorious for casting them in their own image. Prevailing values, moral sentiments, and sexual stereotypes have had a way of creeping into not only our myths and folklore, but our scholarship as well. Cultural biases reflected in nineteenth-century evolutionary theories, for example, now appear quite obvious in hindsight. Ancestral forms prior to the emergence of anatomically modern humans were portrayed as brutish, stupid, oversexed, amoral, anarchistic, and speechless. As the story goes, the threshold of humanity was not crossed until when, thanks to the naturally diminutive libido and greater religiosity of females, males were roped into assuming the responsibilities of family heads, breadwinners, and protectors of dependent consorts and offspring. Thus, ancestral human temperaments and the pair-bonded family unit were perceived as not only mirroring, but rationalizing the Victorian ideal.

Similar elements, however, have also found their way into more recent portraits of Paleolithic social life, along with the notion that these elements were genetically imprinted in the ancient past. This book will examine the extent to which Western European cultural bias has colored our perception of male and female natures, intersexual relationships, the evolution of sociality, the diversity of Paleolithic adaptations, and the antecedents of human kinship systems. In the process, the reader will be challenged to consider alternative viewpoints and interpretations of existing data.

Humans as Chameleons

The prevailing theme in the following chapters is that what set hominins off on a separate evolutionary trajectory—what made us human—was the ability to flex our reproductive and social strategies in response to stochastic conditions. This perspective directly contradicts prevailing monotypic models of Paleolithic life that rely on the genetic imprinting of trait clusters born in Plio-Pleistocene hunting economies. If there is such a thing as a human biogram, it is not a staid template that was perfected for all time in a single ancient biome, but rather the capacity of evolving hominins for plasticity—the ability to tailor their behaviors and adaptations to meet the challenges of changing environmental conditions. This represents a fundamental shift in our perception of human nature and of the hominin evolutionary journey.

This book will argue that human lineages evolved in dynamic mosaic landscapes that selected for flexible rather than rigid adaptive responses. It explores new cross-disciplinary research that links the capacity for behavioral plasticity to critical changes in the structure and organization of the primate brain. Unlike contemporary apes, such as chimpanzees, early humans were equipped with a set of both hardwired genetic codes and "soft inheritance" rules—social DNA—that provided not one, but a menu of standard options for the organization of reproductive and socioeconomic life. What emerges from this discussion is a model of Paleolithic society that challenges prevailing theory on issues such as ancient diet, group size, encephalization, labor division, mating behaviors, and variable systems of kin affiliation. This discussion is undertaken from the perspective of multilevel selection, which addresses how the flexible organization of group life has structured the distribution of fitness-related resources through time.

Chapter 1 begins this journey at the beginning, namely, with conception and an exploration of how differing perspectives on human reproductive biology have influenced past and current theory about male and female natures and their strategies for inclusive fitness. A salient issue is whether the sexes are viewed as pursuing their fitness through cooperative partnerships, or at one another's expense.

Chapter 2 considers how our mammalian origins and multilevel selective pressures may have shaped the evolution of proto-human family groups. Prevailing androcentric models of primeval families are critically evaluated. These theories are then contrasted with alternative perspectives based on the matricentric family, cooperative breeding, and the optimization of male and female fitness within the framework of multimale-multifemale groups.

Chapter 3 addresses the emergence of our genus in the late Pliocene, and considers recent data on the paleoecological conditions thought to have played a role in the evolution and geographic radiation of *Homo erectus*. Contrasting theories on the relative importance of terrestrial and aquatic fauna are examined, and an alternative evolutionary scenario offered that links dietary, morphological, and life history changes in early humans with fundamental shifts in female subsistence and reproductive strategies.

Chapter 4 explores more fully the potential range of dietary protein available to hominins throughout the Pleistocene, and the extent to which assumed reliance on the hunting of mammalian herbivores has influenced the reconstruction of Paleolithic economic and social life. In short, diverse biomes translate into diverse adaptations. Dietary breadth is regarded as a key issue for paleontological questions involving all hominin lineages, including the geographic expansions of *Homo erectus* and anatomically modern humans, and the ultimate fate of Neanderthals.

Chapter 5 pauses to consider signature traits that have characterized the hominin experience through time—the essential qualities that define and bind together both ancient and modern humans. Featured topics include opportunistic omnivory, spatiotemporal awareness, mating patterns, behavioral plasticity, intelligence, social demography, and changes in energetics. This chapter serves as a reminder of the qualities that made hominins unique among primates, and that separate us from other ancient and contemporary nonhuman apes.

Chapter 6 returns to the role played by kinship in structuring human adaptations through time. This chapter explores how kinship has been historically portrayed in anthropological theory and how notions about ancient subsistence patterns, innate dominance, inclusive fitness, and biobehavioral traits shared with contemporary apes have biased our perception of Paleolithic social life. Prevailing theories are reviewed and critiqued, and new genomic evidence is introduced that sheds light on the potential nature of early hominin social groups.

Finally, Chapter 7 addresses the long-standing debate among anthropologists and sociobiologists on the antecedents of matrilineal, patrilineal, and bilateral kinship systems. Kinship is examined as a technology for human niche construction that has allowed humans to manage the two basic elements of fitness—food and sex—by structuring their mating relationships and their social groups in a manner that optimized the recovery of energy and other fitness-related resources in a given ecological setting. The author's perspective on factors that select for uterine and agnatic organization is explored by noting their distinct architectures for structuring reproductive, labor, and political groups in relation to available resources. The chapter explores both the origins and resiliency of matrilineal and patrilineal systems and how these variable strategies for niche construction have responded to change in the post-Neolithic era.

This book will be guided by the initial assumptions presented here. Its story on the origins and nature of human sociality blends mainstream theory and empirical data with some nuanced plot twists. To the extent that its conclusions challenge popular notions about our evolutionary past, the reader is reminded that this endeavor naturally summons a number of theanthropic questions, the answers to which *no one really knows for sure.*

Perspectives on Anisogamy

The casual reader may wonder why a book devoted to the topic of human social origins would begin with a discussion of human reproductive physiology. The simple answer is that all past and present theories about human nature and social evolution make certain assumptions about the significance of male and female reproductive behaviors and how they have helped to shape the biocultural trajectory of our species. In considering these theories, therefore, it is essential to gain an understanding of where their proponents stand on the fundamental issue of male and female reproductive strategies and their effect on human social life over time. This chapter will explore the basics of human anisogamy and discuss how its significance is perceived in both historic and contemporary schools of thought.

Size Matters

Anisogamy, simply defined, is sexual reproduction involving the fusion of gametes of different size or form. In humans, a large, slow-moving egg cell (ovum) is fertilized by a tiny, highly mobile sperm cell (spermatozoon). Egg cells, which are over eighty-five thousand times larger than sperm cells, contain DNA, mitochondria, nutrients, and the resources essential to support new life. Females are born with about 2 million egg cells, only four to five hundred of which will ripen (one each month) over a fertile lifespan, with the remainder degenerating over time. Sperm cells also contain DNA and mitochondria but no nutrients, and are instead optimized for mobility. Gamete production in males is continuous, with about 100 million sperm cells generated daily, or an estimated 2 trillion over a lifetime.

Anisogamy, by its very nature, creates disparate but complementary reproductive roles and strategies. Females produce a small number of nutrient-rich eggs that emit pheromones to attract male sperm for fertilization. Males, in contrast, produce huge numbers of sperm cells that must vie with one another in their typically ill-fated race to unite with an available egg. Since the demand for eggs is greater than the supply, ova enjoy much better odds of passing along their DNA to the next generation than do individual spermatozoa. The greater reproductive investments required of females by gestation and lactation also have the effect of delaying ovulation, thereby further restricting the supply of eggs potentially available to sperm for fertilization. These factors combine to make females the primary limiting resource in male reproductive success.

The traditional textbook version of how the fusion of ova and spermatozoa occurs typically begins with the egg being swept up into the fallopian tube and drifting toward the uterus. Sperm cells released in the vagina move through the female genital tract and race toward the large and essentially dormant egg. At the climax of this intense and somewhat perilous journey, the first among the thousands of sperm to arrive, successfully attack, and forcefully breach the egg's zona or protective barrier wins, and fertilization occurs. This scenario thus paints female gametes as the passive recipients of male competition and penetration, and male gametes as active contestants and penetrators of the coveted target.

Perception of Conception

In 1991, Emily Martin (no relation) wrote an enlightened and entertaining article entitled "The Egg and the Sperm: How Science Has Constructed a Romance Based on Stereotypical Male-Female Roles." Its focus is on how male and female behavioral stereotypes have become infused into the imagery of conception. How could the description of a relatively straightforward biological process be influenced by cultural bias? Martin's paper calls attention to the findings of a Johns Hopkins University study that fundamentally revised the science on how fertilization actually occurs. Whereas it was formerly thought that sperm were the aggressors that burrowed through the egg's zona by the force of their tails, the study discovered that eggs actually capture sperm cells, adhering to them firmly and forcing the head of the sperm to lie flat against the surface of the zona. Successful conception requires a unique and active partnership between the gametes:

> The trapped sperm continues to wiggle ineffectually side to side. The mechanical force of its tail is so weak that a sperm cannot break even one chemical bond. This is where the digestive enzymes released by the sperm come in. If they start to soften the zona just at the tip of the sperm and the sides remain stuck, then the weak, flailing sperm can get oriented in the right direction and make it through the zona—provided that its bonds to the zona dissolve as it moves in. (Martin 1991: 493)

Martin's point is that even in the seemingly objective context of reproductive physiology, biological facts can be construed in cultural terms. Thus, female gametes have been typically characterized as slow-moving, passive, dependent cells waiting to be "deflowered" and seeded with life, while male gametes are aggressive, active, and competitive cells bent on forceful penetration of the hapless egg. The fact that eggs play a very active and complementary role in facilitating fertilization by trapping sperm cells and preventing their escape—that eggs and sperm are mutually active partners in the business of conception—was an idea unexplored by biologists for decades. Instead, eggs and sperm in scientific discourse took on the idealized personalities of females and males in the larger society.[1]

Reproductive Roles and Social Origins Theory

Before the development of modern genetic science, theorists speculating on the evolution of early human sociality relied principally on observations of animal reproductive behavior, along with conclusions drawn from their own cultural experience (including sacred creation myths). The prevailing Victorian viewpoint was that perceived differences in male-female sexuality were linked to innate behavioral traits. Males were naturally sexual, aggressive, competitive, and cognitively superior. Females, in contrast, were both defined and limited by their maternal reproductive functions. As such, they were typically characterized as naturally asexual, passive, nurturing, noncompetitive, and also less capable of complex thought. Several nineteenth-century scholars proposed that these disparate male and female traits had consequences for the way that human society evolved. The abbreviated version of such theories is that, thanks to the alleged sexual restraint of women, early humans emerged from an original state of promiscuity to one of ancient matriarchy, in which society was organized around the obvious biological linkage between females and their offspring. This evolutionary stage was eventually overthrown by males, who established domestic units and enjoyed sociopolitical dominance from that point forward.[2]

Hindsight being 20/20, such notions about innate male and female aptitudes and universal evolutionary stages are now typically dismissed as examples of both gender stereotyping and immature science. But as with the story of the egg and the sperm, cultural bias continues to color our perception of human nature and societal origins. Although presented with a different veneer, behavioral stereotypes rooted in the perceived consequences of anisogamy have been amazingly persistent. International conferences and symposia held in the late 1950s and 1960s framed a model of ancient Paleolithic social life that equated the advent of hunting by males with the origin of human culture itself. As in Victorian times, males were portrayed as sexually dominant and competitive, but as presumably bridling this enthusiasm for the camaraderie of the hunt. This model assigned males active roles as family heads and providers to dependent consorts and offspring, and credited them with the defining achievements in human evolution, such as the invention of tool making.[3] Females, in contrast, were cast in a largely peripheral role, serving as reproductive vessels and performing perfunctory domestic tasks of limited economic value. They were, in essence, silent partners in the evolutionary saga—dependents and nonproducers who traded sexual favors and reproductive functions for economic support.[4]

A subsequent spin-off of this model, the male-bonding hypothesis, gained some popularity in the early 1970s.[5] Its basic thesis was that males are biologically predisposed toward cooperative activities, whereas females lack such built-in genetic codes. Proponents argued that this sexual divide originated from selective pressures that favored bonding among males, initially for group defense and eventually for the predation of the hunt. The model went further, linking such activities to cortical expansion in ancient humans—a consequence of subordinating male sexual competition to the interpersonal cooperation and emotional control necessitated by cooperative hunting. Opportunities for social and intellectual advancement provided by the hunt were allegedly unavailable to the other half of the species due to their reproductive responsibilities and assumed temperament. Females, allegedly burdened by serial pregnancies, inferior physical and cognitive abilities, and swings in "emotional tonus," were not only presumed unfit for the hunt, but to pose a disruptive influence by stimulating male competition for sex. On this basis, the theory concluded that natural selection has favored females who acquiesce to their basic reproductive and domestic roles.

A new chapter on theories of human social origins was introduced in 1975 with the publication of E. O. Wilson's *Sociobiology*. The sociobiology movement spurred significant new scientific research that has led to a re-examination of assumptions about the evolutionary consequences of human anisogamy. Unanimity of opinion among its followers, however, has been elusive.[6] One sociobiological perspective, exemplified by the selfish-gene theory, is that the disparate reproductive strategies of the sexes are inherently conflicting and are played out in an atmosphere of male self-interest, parasitism, and deceit. Other sociobiological theories, in contrast, recognize differences in male-female reproductive strategies but seek to understand how the fitness of both sexes and that of group members is facilitated through the evolution of social behaviors based on cooperative, reciprocal relationships. These contrasting viewpoints are briefly outlined here, and their relevance for theories on human family origins are explored more fully in the next two chapters.

Exploitation and Parasitism

Perhaps the most well-known work linking anisogamy to male social dominance is Dawkins's *The Selfish Gene* (1976). It advances the notion that genes tailor their replication strategies to fit the bodies in which they find themselves and that these opportunities differ for male and female bodies. For Dawkins, it all comes down to anisogamy — all social differences between the sexes can be traced to the fact that sperm cells are smaller and more numerous than eggs:

> Sperms and eggs . . . contribute equal numbers of genes, but eggs contribute far more in the way of food reserves: indeed, sperms make no contribution at all and are simply concerned with transporting their genes as fast as possible to an egg. At the moment of conception, therefore, the father has invested less than his fair share (i.e. 50 percent) of resources in the offspring. Since each sperm is so tiny, a male can afford to make many millions of them every day. This means that he is potentially able to beget a very large number of children in a very short period of time, using different females. This therefore places a limit on the number of children a female can have, but the number of children a male can have is virtually unlimited. *Female exploitation begins here.* (Dawkins 1976: 141–42, emphasis added)

The theory is that each reproductive partner has the goal of maximizing the number of their surviving offspring, but that females are at a disadvantage due to their greater parental investment. According to Dawkins, two principal "counter-ploys" may be pursued by females to reduce their exploitation by males. The first, dubbed the "domestic-bliss" strategy, envisions females as feigning "coyness" in order to assess potential mates in advance for favorable attributes (i.e., fidelity and domestic assistance) and to demand material investments prior to copulation. The second is referred to as the "he-man" strategy, where females, in lieu of securing parental assistance for their offspring, settle for trying to select mates that appear to have "good genes."

These strategies notwithstanding, Dawkins's scheme offers females little hope for escape from exploitation by promiscuous males, who are

28 • Social DNA

Figure 1.1. Selfish-gene theory and the origin of female exploitation. By Drew Fagan.

portrayed as naturally self-serving, devious, and irresponsible. Saddled as such with their "egg yokes," females are equipped with few liberating tools beyond withholding sexual favors, clandestine entrapment of reluctant progenitors, or luck of the draw.

Dawkins's selfish-gene hypothesis is taken to its logical extreme by van den Berghe (1979), who adopts it as a template for understanding the structure of human social groups, past and present. He proposes that coyness and slower erotic arousal have evolved as female strategies to counter male seduction and to ensnare men into monogamous or pair-bonded relationships. Males, in contrast, are naturally promiscuous, and seek to parasitize women for their reproductive potential. Male dominance, patrilineal descent groups, and polygyny are seen as natural outcomes of the desire of men to secure the reproductive power of women, assure paternity, increase the number of offspring, and maximize their inclusive fitness by passing property on to their sons. Matrilineal social groups, while recognized, are viewed as rare and aberrant, arising where paternity is made less certain by a high incidence of adultery, divorce, and the "cuckolding" of absentee males. In these situations, it is argued, parasitism by males is simply transferred to their brothers-in-law for the benefit of their uterine nephews, with whom they share genes through their sisters. According to van den Berghe (1979: 108): "In both types of society, women are dominated by men who set norms in an attempt to control women's reproductive behavior for maximum male fitness."

The picture that emerges from such selfish-gene models, then, is that anisogamy has preprogrammed the sexes for asymmetrical and parasitic reproductive roles in the evolutionary drama. Males are typecast as the sexual protagonists and females as the reticent objects of their exploitation. Similar conclusions on the proclivities and copulatory appetites of the sexes were reached by Daly and Wilson (1978) and by Symons (1979).[7]

Such theories generated immediate controversy and academic debate within the field of evolutionary biology. Is our species the product of a grand genomic zero-sum game? Were the selective pressures that favored cooperative behaviors and encephalization in evolving hominins exerted largely on males? Are female primates naturally passive, noncompetitive, and asexual? Continuing research over the ensuing decades has shed new light on these questions, and has challenged previous notions about both human sexuality and human societal origins.

Complementarity and Cooperation

An alternative perspective on the social consequences of anisogamy is that selective pressures favored the evolution of disparate but *complementary* behaviors among the sexes within the context of cooperative breeding communities. This theoretical framework assumes that *both* males and females are active players, each pursuing their reproductive strategies sexually, politically, and economically, and further, that their mutual reproductive success is enhanced by the creation of collaborative and overlapping alliances.

By the mid 1970s, androcentric theories on human societal origins were being increasingly challenged by the women's movement, and female scholars began to re-examine prevailing assumptions in the fields of anthropology and primatology.[8] Sarah Hrdy, in her seminal work *The Woman That Never Evolved* (1981), questioned the notion that natural selection had exerted its influence primarily or exclusively on males. The book presented data that challenged existing stereotypes of primate behaviors, which had previously relied largely on observations of male dominance hierarchies and differential male access to estrus females.

Women primatologists took to the field, placing primary research emphasis for the first time on the sexual and social lives of females. The results of these studies literally transformed our understanding of primate breeding systems (Small 1984, 1993).

What researchers discovered is that female primates are far from the coy, asexual, demure, and parasitized creatures some portrayed them to be. Instead, they are often sexually aggressive, actively seeking the novelty and variety of multiple partners while utilizing sex to create strategic alliances that further their reproductive success. Females are also often highly competitive, commonly organizing themselves into stable hierarchies in which status and territories are passed from mother to daughter. These hierarchies may impact access to food resources and, because they have greater stability and permanence than male mating hierarchies, can play a critical role determining the overall social structure of the group. In short, female primates are deliberate and active participants in the game of reproductive success. The strategies of males and females are distinct but interdependent, and rely on an intricate web of same-sex and opposite-sex alliances. What these primate data suggest is that natural selection has exerted its pressures on the sexes from ancient times, both on individuals and on groups within the context of complex social networks that bind the members of a breeding community.

The perception of anisogamy pursued in this book is consistent with this view. Just as the joining of the sperm and egg at conception requires a mutually active partnership, so the evolution of human society required the successful integration of male and female reproductive strategies. In the words of Christopher Ryan and Cacilda Jetha: "There's no denying that men and women are different, but we're hardly different species or from different planets or designed to torment one another. In fact, the interlocking nature of our differences testifies to our profound mutuality" (2010: 270).

Models of why and how intersexual partnerships evolved need not be based on notions of conflict, parasitism, dominance, gender stereotypes, or zero-sum games. Rather, human advancement relied more on brains than brawn—that is, on the liberation of males and females from primitive limbic system responses and the corresponding ability to modify their behaviors and structure their alliances in ways that furthered their respective reproductive goals. The next chapter challenges prevailing theories on the nature of proto-human society and offers an alternative model of how ancient multimale-multifemale communities evolved on the basis of cooperative intra- and intersexual alliance.

 2

First Families

Primates, past and present, share anisogamous reproduction with many other social mammals. Anisogamy equips and predisposes the sexes to pursue different reproductive strategies. Generally speaking, inclusive fitness for males is measured by the quantity of offspring they produce, thereby increasing their genetic contribution to subsequent generations. As such, the key resource affecting their reproductive success is access to fertile females. In contrast, inclusive fitness for females is shaped by biological limits on the number of potential lifetime conceptions they may achieve and by the parental investment required to successfully nurture their resultant offspring from zygote to maturity. The key fitness resource for females is thus access to food sufficient to meet the nutritional requirements of themselves and their issue while keeping their offspring out of harm's way. It is the interaction of these disparate male and female reproductive strategies that shapes the social life of a primate group.

As noted in the previous chapter, scholars have placed reliance on the perceived consequences of anisogamous reproduction in reconstructing the social life of ancestral hominins. Historically, these theories have been based largely on observations of male primate mating patterns and have attached particular significance to the subordination and control of females. Not surprisingly, the resultant models have prescribed male-centered social organization as a founding and pervasive feature of our evolutionary past. More recently, attention has focused on female intra- and intersexual alliances and the way in which female competition, bonding, cooperative breeding (alloparenting), and access to resources influence social group structure, including the distribution of males. This alternative perspective has provided new insights into the complex interplay of male and female reproductive strategies, and the dynamic role played by ecological factors in primate sex and subsistence activity. The findings of such integrative studies behoove us to reconsider the epigenetic nature of proto-human social groups.

Framing the Primeval Family

Members of the Family Hominindae[1] shared a common family tree in the Late Miocene period, with gorillas branching off from the ancestral line between 8 and 10 ma, and chimpanzees about 5 ma. The last common ancestor (LCA) of our closest living relatives, the chimpanzees and bonobos, is thought to have occupied habitats on the forest fringe and to have been bipedal and at least semiterrestrial. This LCA gave rise to a variety of proto-human bipedal apes during the Pliocene, such as *Ardipithecus ramidus* (4.4 to 4.3 ma), along with a radiation of australopithecine taxa approximately 4 to 2 ma. Australopithecines were tool-using foragers that occupied more open, mosaic habitats, and have frequently been identified as the immediate predecessors of the genus *Homo*, although some investigators point to earlier, less morphologically dimorphic forms.[2]

This chapter explores theories on the social nature of our LCA—on primeval traits that may have been genetically imprinted at the time of the *Pan-Homo* split and that continue to be reflected in the behaviors of contemporary apes and humans. Reconstructions of proto-human social organization are necessarily speculative and rest on a number of key assumptions that may not always be explicitly stated by the architects of such theories. These foundational elements must therefore be called out and carefully examined. Among them are the following:

Plio-Pleistocene Paleoenvironment

Fossil, geologic, and paleoclimatic information suggests that hominin evolution took place during the Late Pliocene in tandem with the gradual displacement of tropical forest by more open mosaic and savanna habitats. Proto-human apes adapted to this dynamic landscape by expanding the scope of their food-getting activities and by restructuring their corporate subsistence groups. A key element in social origins theories, therefore, concerns the nature of the ecological setting or settings in which such developments are assumed to have taken place. As discussed more fully in chapter 3, academic debate has centered on whether these habitats were open grasslands or forest margins, arid expanses or well-watered lakeshores and river deltas, or perhaps all of the above—a variable landscape of shifting vegetative and climatic regimes.

Ecological variation affected the range of flora and fauna potentially available to proto-humans and early hominins, as well as the type and relative importance of subsistence activities pursued by females and males. As noted later in this chapter, the density and distribution of food resources within a given niche is known to influence the prevailing spatial distribution of female and male primates, and hence is a potential determining factor in the social structure of their communities. A theorist's assumptions about the primeval ecological setting in which the LCA lived and evolved thus sets the stage for conclusions about the selective factors and adaptations that shaped proto-human social life.

Natures and Reproductive Strategies of the Sexes

Theories on the nature of proto-human society make a number of assumptions about ancient primate mating patterns, as well as the motivations and temperaments of the actors. Were intersexual relationships permanent or casual? Exclusive or nonexclusive? Egalitarian or inequitable? Coercive or voluntary? Parasitic or symbiotic? What strategies did males and females utilize to pursue their inclusive fitness? Were these strategies competitive or complementary? As noted in the previous chapter, answers to these questions reflect theorist biases about the genetic basis of male and female behaviors and the extent to which the sexes played active or passive roles in the structuring of proto- and early human social groups.

Pair Bonding, Paternity, and Provisioning

By virtue of our mammalian nature, the genetic relationships and kinship bonds established between mothers and offspring are a given. The resultant *matricentric* family units are the essential building blocks of all primate social groups. Unlike most contemporary primates, however, humans characteristically rely on the variable socioeconomic investments of both sexes for reproduction and survival. Theories on human family origins must therefore account for how and when males came to be joined to matricentric units and integrated into cohesive local communities. Did males establish themselves as family heads by forcibly herding one or more females into harems for the purpose of exclusive breeding? Or, alternatively, were unattached males selected by females to join existing female-bonded groups on the basis of their positive attributes for mating, parental care, and defensive support?

Earliest theories on social origins embraced the notion that males and females emerged from an original state of "primitive promiscuity" to bond with one another in more permanent reproductive relationships. Male-female pair bonding and the alleged universality of the nuclear family is perhaps one of the most pervasive concepts in the social sciences. The domestic unit of father, mother, and offspring has been long proposed as the seminal and enduring foundation of human social groups. Indeed, artists' images of early hominins typically portray them as unitary conjugal pairs, strolling arm-in-arm across a barren plain or sharing a crude encampment in mock display of distinct domestic and provisioning roles.

Arguments made in support of ancient male-female pair bonding have both sexual and economic components. The sexual rationale has been traditionally grounded in male mating behavior, namely, that pair bonding evolved to accommodate male sexual aptitudes and appetites. Sahlins (1960), for example, argued early on that conjugal unions provided a mechanism for reducing sexual competition between males for fertile females, thereby freeing them for cooperative pursuits. Alexander and Noonan (1979), Lovejoy (1981), Wilson (2012: 253), and others have subsequently supported the notion that the continuous sexual receptivity and concealed ovulation (*ovulatory crypsis*) of early hominin females lured males into pairing arrangements by providing both copulatory outlets and assurances of paternity. Symons (1979) went even further, suggesting that the female sexual orgasm itself is a uniquely human trait that evolved solely because it was adaptive for males.[3]

The economic rationale for ancient pair bonding has been based on the premise of female subsistence dependency and the provisioning of mothers and offspring by male consorts. The "sexual contract," wherein females exchange sexual favors for male support, has its roots in hunting and male-bonding theories that link early terrestrialism and the provision of meat by males to dependent families as a cornerstone of hominin development. In such models, females are portrayed as performing domestic and child-rearing tasks at a home base, while males wander off to find high-quality foods that ensure family survival. By juxtaposing females and males in domestic versus provider roles, formation of the nuclear family unit is also credited with the origin of the division of labor by sex.

In addressing the structure of male-female relationships, therefore, theorists make a number of assumptions about the motivations and trade-offs for both sexes, and how different mating arrangements relate to the spatial distribution and subsistence basis of family units. As discussed in this and subsequent chapters, pair bonding and male provisioning continue to appear as centerpieces in sociobiological reconstructions of proto-human society. Models supported in this book, in contrast, assign a more prominent evolutionary role to the matricentric family, female provisioning, female mating choices, and the influence of ecological factors on the architecture of human social groups.

Analogs and Avatars

When looking for clues on how proto-human and early hominin societies may have been structured, theorists typically turn to two primary sources—contemporary nonhuman primates and simple hunter-gatherer populations. By comparing the social behaviors of living primates with the foraging patterns, sex roles, and social groups of surviving preagricultural peoples, investigators seek to identify common traits that may be explained on the basis of either convergent evolution or shared descent from a common ancestor. In noting how such interspecific comparisons can be used, Jolly (2009) drew the distinction between *analogs* (shared traits that may reflect the selective factors and similar functional adaptations from which they were derived), and *avatars* (living animals thought to sufficiently resemble extinct ones to suffice as models of ancient species). Chapais (2011, 2014) made a similar distinction between *homoplasies* (interspecific traits derived from similar natural selective pressures) and *homologies* (interspecific traits rooted in common descent).

As discussed below, reconstructions of our Pliocene LCA differ in the relative importance they assign to biological and ecological factors in shaping the trajectory of human evolution. Some theorists utilize interspecific comparative studies to highlight parallel traits in humans and selected primate avatars, such as chimpanzees or baboons, suggesting that such shared traits are indicative of phylogenetically derived behaviors. Analogs, in contrast, are used as referents in origins theories to understand the social behavior of primates as a function of both natural selection and adaptation to the environments in which they and their ancient forebears lived. While both are important analytical tools, the distinction between analogs and avatars is sometimes blurred in evolutionary models. Apes and humans share a common ancestor and hence some common traits, but all contemporary primates—*Homo sapiens* included—are the product of adaptations to comparatively recent conditions that may or may not reflect conditions in the ancient past.

Ancestral Male Kin Group Hypotheses

The notion that ancient society was necessarily structured on male kinship principles has a long history in Western European culture, and continues to represent the prevailing bias today. Historically, it has been based on assumptions of male physical dominance, aggression, sexual prowess, family provisioning through game hunting, and the allegedly unique propensities of males for cooperative activity. The result has been a rather one-dimensional portrait of human social organization spanning the past several million years. This prototype is based on the spatial dispersal of related females at sexual maturity and the local aggregation of related males into cooperative kin-based groups, a pattern known as *male philopatry*. Male-centered theories were prominent in the nineteenth century, and further elaborated in the twentieth-century writings of Steward (1955), Sahlins (1960), Service (1962), Lee and Devore (1968), Tiger (1969), and others. These early models, which often focused on identifying a common basis for human and nonhuman primate social behaviors, have experienced renewed popularity on the heels of the sociobiology movement.

The rationale for ancient male philopatry on which recent theories have relied is grounded in Ghiglieri's (1987) initial sociobiological reconstruction of the *Pan-Homo* ancestor. Ghiglieri cited cross-cultural studies to argue that patterns predominating in contemporary human societies, such as male philopatry, female dispersal, polygyny, and territoriality, have a phylogenetic basis. He noted that these trait clusters, while uncharacteristic of orangutans and gorillas, are present among our closest living relatives, the chimpanzee and bonobo. On that basis, he concluded that the *Pan-Homo* LCA evolved these "rare social behaviors" subsequent to the separation of orangutans and gorillas from the main hominid line. By remaining in their natal territories, Ghiglieri reasoned, clusters of related males could cooperate in their reproductive efforts, cloistering and breeding with multiple females, fostering reciprocal altruism among close kinsmen, and defending females, offspring, and territorial resources from external threat. Ghiglieri's hypothesis that this trait complex has a genetic foundation dating back to the *Pan-Homo* LCA has been widely embraced, and has formed the foundation for parallel theories proposed by Foley and Lee (1989), Rodseth et al. (1991), Chapais (2008, 2011, 2014), Lovejoy (2009), and Greuter, Chapais, and Zinner (2012). Dissected further, ancestral male kin group hypotheses share the following basic elements and assumptions (Table 2.1).

Primeval Environment and Subsistence

A platform assumption of ancestral male kin group hypotheses is that the defining environment in which our *Pan-Homo* ancestors lived consisted of arid or semi-arid savanna grasslands with widely dispersed, patchy, and seasonally limited resources. This type of setting translated into large subsistence ranges, low population densities, and intergroup feeding competition. Foley and Lee (1989) concluded that such habitats

Table 2.1. Principal assumptions about *Pan-Homo* social life in androcentric models.

Primeval Habitat	Open savanna grasslands
Food Distribution	Widely dispersed and seasonally limited resources; large subsistence ranges; intergroup feeding competition
Demography	Small, nomadic territorial bands
Philopatry	Male retention and female dispersal for incest avoidance and optimization of male fitness; exchange of females for intergroup alliance
Mating & Family Structure	Aggressive male herding of consorts into one-male units; polygyny, evolving into pair-bonded monogamy
Intersexual Relationships	Male dominance, female subordination; male bodyguarding of females against roving bachelor males; male consort guarding for prevention of female cuckoldry; male provisioning and sexual contract evolving with pair bond
Intrasexual Relationships	Male kin bonding and fraternal interest groups; amalgamation of one-male groups into multilevel social systems; female bonding prevented by male philopatry and constraints on cooperative subsistence activity imposed by scarce, dispersed resources
Avatars/Analogs	Chimpanzee; Hamadryas baboon

would have been incompatible with cooperative female food gathering activities due to resource scarcity and would therefore have militated against the spatial aggregation of related females. Instead, they argued that male kin coalitions would be favored, strengthened by both cooperative hunting and the joint defense of females and young from bands of "roving or intruding males." Greuter et al. (2012) also attributed the development of male-centered groups to sparsely distributed resources, aggressive male herding of consorts, and the protection of females and offspring from the hostile actions of bachelor males. In short, a key assumption of ancestral male kin group models is that the LCA lived in a relatively resource-poor environment that was fraught with inter- and intraspecific danger—circumstances that compelled males to serve as bodyguards and provisioners, and females to abandon their natal families at puberty to seek the shelter and succor that consorts provided.

Sexuality, Mating, and Fitness

Males are the principle protagonists of sexual and mating relationships in all ancestral male kin group models. Indeed, a central premise of such models is that the hominid-to-hominin transition was driven largely by male struggles for reproductive success. Ghiglieri's (1987: 345) statement that male philopatry represents a "major evolutionary leap" is based on the premise that becoming human required the exclusion of female bonding, precisely because such social networks were viewed as contrary to the furtherance of communal male reproductive goals. This scenario paints a universe in which males aggressively seek to monopolize females for their breeding potential, enforce their dispersal from natal groups to avoid incest, and exchange them as commodities to forge alliances and to build multilevel communities with other fraternal interest groups. Female inclusive fitness clearly plays a subsidiary, if not silent, role in this evolutionary drama, with their reproductive success wholly dependent on passivity, acquiescence, subordination, and entry into sexual contracts for male provisioning and protection.

Philopatry and Social Structure

Ancestral male kin group models generally accept the premise that multimale-multifemale breeding communities similar to those of contemporary primates, wherein both males and females have multiple mates, existed at some time in the distant past. The task then becomes one of accounting for the evolution of a distinctly human complex of traits, such as incest avoidance, recognition of paternity, regulation of mating relationships, maintenance of lifelong bilateral kinship bonds, and the tendency to aggregate into multilevel communities. For the architects of androcentric theories, this trait complex flowed naturally from male reproductive and social dominance. In other words, the social evolution of hominids to hominins is marked by the increasing elaboration of male kinship alliances, beginning with the herding of females into polygynous one-male groups, and finally into monogamous pair-bonded relationships. Notions on how this process proceeded varies somewhat from theory to theory, but the end product is the same.

A central tenet of such theories is that male philopatry was already established at the time of the *Pan-Homo* split. Notably, Rodseth et al. (1991) acknowledged that the overwhelming majority of contemporary primates are characterized by *female* rather than male philopatry, and that human social traits such as the incest taboo, exogamy (mating outside one's natal group), pair bonding, and recognition of bilateral kin could have theoretically evolved from either male or female dispersal patterns. They also recognized that the organization of primate communities around related females (matrilocal residence) contributed to the strengthening of cross-kin alliances and to politically stable multikin communities. At the same time, however, female philopatry was perceived by Rodseth and colleagues from a decidedly male perspective as "agnatic sprawl." Based on observations of modern apes such as chimpanzees, they argued that male philopatry creates a tighter "modular" system, wherein intergroup alliances are cemented by the exchange of females who act as "peacekeepers" between volatile groups of male kinsmen.

Chapais (2008) also proposed that the *Pan-Homo* LCA lived in chimpanzee-like multimale-multifemale groups characterized by male philopatry, female dispersal, and male kinship bonds. A more detailed analysis of his reconstruction of human kinship origins is undertaken in the chapter 6 discussion of *Pan*-genesis. Briefly, he argues that the hominin family was derived from this base through the creation of more stable relationships by males, first through polygynous couplings and eventually monogamous pairs. Chapais drew attention to unique features of human social structure, such as the tendency of humans to organize themselves into multilevel (or modular) societies, aggregate into communities of multiple conjugal pairs, practice male or female dispersal at adolescence, and maintain lifetime bilateral and affinal kinship bonds. Pair bonding is credited as the pivotal factor in the evolution of these derived traits, the rationale being that the combination of monogamy and male philopatry allowed the offspring of early hominins to recognize their fathers and male kinsmen and, by extension, the bilateral relationships with their maternal and affinal lines.

Lovejoy (2009) discounted the likelihood of ancient polygynandry (multiple mating by both sexes), and instead proposed that male philopatry and kin selection were "proscribed" by evolving reproductive strategies that favored advanced K-selection[4] and the dispersal of females for purposes of incest avoidance. The separation of related females from their natal group, Lovejoy argued, had the effect of reducing the benefits of female cooperative allocare of offspring. In his view, it therefore encouraged compensatory developments such as male provisioning (including food transport and terrestrial bipedalism), pair bonding, and the sexual contract among consorts. The evolution of female ovulatory crypsis was also seen as a necessary accompaniment to male parental investment in offspring due to its purported role in reducing female "cuckoldry" and sexual competition among males for estrus females. For Lovejoy, male philopatry led to what he termed an "early hominid adaptive suite," consisting of male cooperation for enhanced provisioning, territorial protection, reduction of internal competition, and maintenance of group social cohesion.

This basic ancestral male kinship model was further elaborated by Greuter, Chapais, and Zinner (2012), who credit it with the evolution of multilevel primate societies. They examined the limited number of primate clades that exhibit multilevel social systems (papionins, Asian colobines, and hominins) with an eye to identifying their homologous origins. The authors propose that the "one-male unit" (OMU), consisting of a male and one or more females and young, is the basic building block of primate multilevel systems. The social organization of Hamadryas baboon (*Papio hamadryas*) bands, in which OMUs of related males are nested into "clans," is offered as a model of primate modular societies that may have evolved from the splitting of large polygynandrous multimale-multifemale (mm-mf) groups. Greuter et al. propose that promiscuous mm-mf LCA groups underwent a similar fission process due to resource scarcity and external threat. They theorize that ancestral hominins reorganized themselves into polygynous OMUs to reduce feeding competition and enhance both male protection and reproductive goals ("male guarding hypothesis"). These OMUs subsequently reassembled into bands at shared refugia or home bases where more abundant localized resources served to reduce food competition ("localized resource hypothesis" and "predator avoidance hypothesis"). The second phase of hominin evolution is described as the transition from polygynous OMUs to monogamy. Their theory proposes that movement away from polygyny was precipitated by the establishment of male provisioning, the high cost of supporting multiple females, and a reduction in male sexual competition.

Ancestral male kin group theory thus relies on the construction of a rather elaborate pyramid of hypotheses, all of which are based on debatable assumptions of resource scarcity, intraspecific competition, external threat, and male provisioning. As noted later in this chapter, this model also fails to account for the evolution of multilevel systems among papionin species in which female rather than male philopatry predominates.

Primate Avatars and Analogs

Ancestral male kin group theories rely heavily on the social organization of contemporary chimpanzees as evidence for the antiquity of male philopatry in the hominid family tree, thereby establishing a phylogenetic basis for its expression in subsequent hominin populations. As noted above, Greuter et al. (2012) also pointed to one specific type of baboon, the hamadryas, as a model for how male philopatry is linked to the development of more complex multilevel or modular social groups. In an effort to establish a phylogenetic basis for ancient male philopatry, proponents such as Greuter et al. and Bribiescas, Ellison, and Gray (2012) also looked for evidence in the fossil record by citing the results of a strontium isotope analysis of South African australopithecine teeth. This study, conducted by Copeland et al. (2011), explored the hypothesis that geological substrates of areas occupied by early hominins during the period of tooth mineralization would be reflected in the isotope compositions of tooth enamel, and hence provide a window on ancient landscape use patterns. Fossil teeth from nineteen individuals were sex-typed on the basis of relative size, and their isotopes compared with local and nonlocal substrates. The results of such measurements suggested to researchers that teeth identified as female belonged to individuals more likely to have dispersed from areas in which they spent their early years. For reasons to be discussed below, however, such generalizations about australopithecine social patterns remain highly speculative.

To summarize, the notion that ancestral apes living at the time of the *Pan-Homo* split were organized around localized clusters of related males has been a persistent theme in Western paleoanthropology. Male philopatry and female dispersal are proposed as the natural outcome of male reproductive strategies that were biologically imprinted in the LCA and emergent hominin populations. Localized fraternal interest groups are portrayed as a central organizing element for male cooperative subsistence and defensive activity, and the exchange of females between such groups as a unifying element in the emergence of multilevel organization. While environmental and ecological variables are recognized as limiting factors, such models assign clear priority to the biological origins of hominin social systems.

Shortcomings of Androcentric Models

Models of human social evolution based on a unitary design carry both the benefit of parsimony and the curse of exclusion. That is, the assumption that a single hominin social blueprint prevailed throughout the Pleistocene and beyond belies the diversity witnessed among both human and nonhuman primates. The alternative position taken here is that behavioral plasticity and the ability of hominins to vary their reproductive and social strategies to meet the challenges of changing environmental conditions was integral to their evolutionary success.

Pliocene Ecology Revisited

One of the principal shortcomings of androcentric models is their failure to account for or explain the potential range of diversity among both ancient and contemporary primates. Observations in the wild have documented the ability of primates to adapt to a wide variety of environmental settings and in so doing to modify their social strategies to optimize the exploitation of available resources. This extraordinary plasticity has provided both human and nonhuman primates with the adaptive advantage of varying the spatial distribution of females and males, and hence their core social structures, in accordance with the spatial distribution of foods in a given niche. The establishment of a single monotype for ancient social organization, in essence, requires the parallel establishment of a monotypic environmental setting in which ancient selective pressures are presumed to have held sway.

As discussed in more detail in chapter 3, recent evidence raises questions about the generalized aridity of African landscapes where hominins are thought to have evolved. In short, the paleoecology of these areas in the Late Pliocene appears to have been much more varied than previously supposed. Proto-human apes are now known to have lived in mosaic habitats that contained a dynamic mix of riverine forests, lakeshores, and edaphic grasslands. These environmental settings offered a variety of types and distributions of flora and fauna, the optimal exploitation of which would have shaped ancient social life in different ways.

What if, contrary to the assumptions of the ancestral male kin hypothesis, some primeval habitats were well-watered, were more resource-rich, or offered foods that were distributed in discrete, defensible patches compatible with collaborative female foraging patterns? Under these more variable conditions, it is reasonable to assume that the size and social structure of proto-human primate groups were not monotypic, but enjoyed a range of diversity similar to that of contemporary primates.

Gender Stereotyping and Fitness

A second shortcoming of popular androcentric models is their narrow, preemptory focus on male dominance and male reproductive success. The characteristic behaviors and aptitudes of prehistoric males and females suggested by such models carry forward the old Victorian stereotypes of active and passive sexes. Males are cast in the active roles of family makers, providers, and defenders of female vulnerabilities and virtue, all in pursuit of their own ends. Whether herded into harems or parties to sexual contracts, females in such scenarios are bullied and bartered by males intent on enhancing their reproductive largess, and then left to seek their own fitness under the uncertain mantel of male protection and occasional acts of chivalry.

This narrow perspective on hominin inclusive fitness belies the depth and breadth of sexual and alliance strategies pursued by both sexes among contemporary primates, and that provide an alternative model for ancestral primates as well. Hrdy (1981) was one of the first to challenge assumptions about female sexual passivity and male family provisioning that underlie traditional male philopatric and pair-bonding theories. Field data suggest that female primates increase their inclusive fitness by utilizing an active, promiscuous sexuality to forge relationships with multiple males. According to Hrdy: "What we know about primates suggests that prehominin females embarking on the human enterprise were possessed of an aggressive readiness to engage in both reproductive and nonreproductive liaisons with multiple, but selected males" (1981: 176). By creating a network of possible progenitors, females enhance their reproductive success by enlisting the potential contributions of multiple male consorts to the survival of their offspring. This support may take the form of protecting them from harm by providing defense against predators, refraining from aggressive acts against their infants, or, as among bonobos, simply living in harmony. Small (1984, 1993) also documented the active pursuit of multiple mates by female primates, even in monogamous species and harem groups. The most common primate social organization consists of multimale-multifemale groups with mutual mating access, or polygynandry. Monogamy among primates is exceptional, documented only for a minority of prosimians and monkeys.[5] Male-female pair bonds or "friendships" do occasionally form among several primate species, but they are usually temporary in nature and do not involve food provisioning.

Female reproductive success is critically linked to the acquisition of high-quality foods, access to which is often structured by intrasexual competition and alliances rather than by sexual contracts with males. Most primate species are based on a corps of resident females, organized into matricentric rather than conjugal family units whose memberships enjoy ranked, sometimes hereditary privileges to exploit local food resources. In some primate species, such as gorillas, female matricentric units join polygynous collectives headed by a dominant male, or, as among chimpanzees, forage independently within the atomistic boundaries of related-male kin groups. In each case, females take the initiative to establish creative sexual and social alliances with resident males and females that optimize their access to resources and ensure the well-being and security of their offspring.

As noted by Linton (1971), it is just as likely that the evolutionary origins of provisioning and the division of labor occurred first within these matricentric family units rather than between sexual partners. In other words, individuals who shared food were more likely to do so with those who shared food with them, namely, their mothers and siblings. Similarly, given the prominence among primates of female philopatry and multigenerational matricentric kin groups, there seems to be no need to propose paternal recognition and investment as prerequisites for the extended nurturance of offspring. These same matricentric kinship units, with their potential for ranking collaborative activities by both age and sex, have greater permanence than mating relationships or male dominance hierarchies, and likely also provided the seminal structure for extended caretaking and provisioning.

When the inclusive fitness concept is applied to primates of *both* sexes, the counterbalancing effect of female reproductive strategies and the politics of inter- and intrasexual alliances are highlighted. Since the reproductive strategies of females vary with environmental conditions, and since male strategies tend to track those of sexually available females, this broader perspective raises important questions about the range and diversity of ancestral primate social organization and the extent to which behavioral plasticity rather than encoded responses provided a selective advantage.

Philopatry and Human Trait Complexes

As acknowledged by Ghiglieri (1987: 344), male philopatry and female dispersal predominates in only 5 percent of contemporary nonhuman primates. Indeed, the majority pattern among primates generally is polygynandrous mating, male dispersal, female philopatry, and the bonding of related females, the social units of which sometimes coalesce into multigenerational kinship units. The question for androcentric models, then, is why male philopatry is relatively rare among nonhuman primates. In Chapais's (2014) proposed classification of primate phylogenetic traits, recognition of kin relationships among mothers, offspring, and siblings is considered a "primitive" trait—one that presumably arises from an awareness of obvious biological relationships. However, more extended bonds among matrilineal kin, he argues, were not present among our *Pan-Homo* ancestors, since the requisite pattern of female philopatry was allegedly absent. By this somewhat circular argument, Chapais proposes that male philopatry is a "concestral" or *Pan-Homo* trait by virtue of its homologous presence among chimpanzees and humans. It was the local aggregation of related males, he proposes, that laid the foundation for hominid "deep social structure."

An alternative view is that the monotypic pattern of male retention and female dispersal is not a necessary prerequisite or precondition for the evolution of human social organization. Other viable pathways exist for the emergence of traits such as incest avoidance, the maintenance of long-term relationships with dispersing siblings and offspring, the establishment and maintenance of intergroup alliances, and development of multilevel social structures.

Theories on the origin of incest prohibitions are a case in point. Jolly's (2009) "frontier hypothesis" links exogamy to male philopatry with the argument that males were encouraged to remain in their natal territories because the rapid range expansion of proto-humans would have reduced the number of neighboring groups containing suitable female mates for dispersing males. The counterargument, of course, is that the search for suitable unrelated mates would seem to pose a similar challenge for females. The dispersal of individuals from their natal kin group at breeding age, whether male or female, is exogamic by definition. Thus, either philopatric pattern serves the purpose of reducing the likelihood of incest. Moreover, where female philopatry prevails, primates tend to aggregate into multikin communities, thereby allowing males to find suitable mates in close spatial proximity to their natal group. As acknowledged by Jolly (2009: 196): "Individuals presumably moved from one group to another in search of opportunities to find mates and congenial social contexts, but (although chimpanzee-like female dispersal and male philopatry are often assumed) we really have no idea whether early human dispersal affected males, females, or both."

A second distinctively human characteristic attributed to male philopatry is the establishment of enduring kinship bonds. Ancestral male kin group hypotheses imply that maintenance of consanguineal kin ties with dispersing members relies on the recognition of bilateral kinship and what Rodseth et al. (1991: 240) refer to as the "release from proximity." Such bilateral kinship principles are purported to emerge from monogamous pair bonding and the assurances of biological paternity provided by exclusive sexual partnerships. Thus, in the case of male retention, both parents are recognized by offspring and categorical distinctions are made between "father's people" and "mother's people." Chapais (2014) theorizes that since the recognition of father-child relationships is lacking among chimpanzees, the evolution of patrilineal kinship groups must have occurred sometime after the *Pan-Homo* split.

An opposing viewpoint is that the maintenance of lifelong consanguineal kinship ties and the evolution of kinship networks with non-natal, nonlocal groups has no necessary linkage to pair bonding or to the recognition of biological paternity. With female philopatry, the essential social links are between mothers, offspring, and siblings—between groups of matrilineal kinsmen rather than sexual partners. The relevant classificatory distinctions thus become "mother's people" versus "peoples with whom brothers or mother's brothers reside" and "peoples with whom I can mate." With this type of social structure, assurance of paternity is neither a burning issue nor a deterrent to the establishment of complex cross-kin alliances. For males in female-centered primate groups, recognition of biological paternity takes a back seat to fealty to one's mother's kinship group and to subsequent residence in a non-natal social group that provides reproductive access to one or more fertile females.

Similarly, the localization of fraternal interest groups is not a necessary prerequisite for the evolution of multilevel social networks. Advocates of the ancestral male kinship hypothesis acknowledge that intergroup alliances may be established and maintained with either male- or female-centered social systems. With female philopatry, the dispersal of related males promotes intergroup alliances by establishing reciprocal relationships between local matrilineages. Matrilocal societies tend to be organized into endogamous (in-marrying) multikin communities, in which the interlineage exchange of males fosters stable internal relations, while uniting them at the supralineage or community level for common interest purposes. Rodseth et al. (2012: 231), in contrast, see the dispersal of male kinsmen—or what they term "agnatic sprawl"—as a deficit that is "remedied" by patrilocality. In their scheme, male-centered interlineage alliance systems, cemented by the exchange of female kin, emerge as ostensibly more cohesive than their matrilineal counterparts due to the central pan-evolutionary role assigned to male bonding and cooperation. As discussed at greater length in chapter 7, however, societies that localize rather than disperse related males have been documented as having the highest levels of feuding, internal warfare, and political instability (Otterbein and Otterbein 1965; Otterbein 1968).

Selective Prototypes

Theorists advocating for ancient male philopatry have consistently pointed to one of our closest living relatives, the chimpanzee, as an exemplar of ancient hominid social life from which a complex of distinctly human traits evolved. While only 5 percent of nonhuman primates rely on male philopatry as an organizing principle, the predominance of this social pattern among chimpanzees is often pointed to as evidence for evolutionary continuity in our family tree. This opinion, however, is not unanimously held. Lovejoy (2009), for example, while embracing the notion of ancient male philopatry, discounts the use of both contemporary apes and australopithecines as avatars for the LCA due to their retention of robust dimorphic features such as large honing canines and their primarily frugivorous or folivorous diets. Instead, he argues that our hominin ancestors more closely resembled ancient Pliocene apes such as *Ardipithecus ramidus* that were bipedal, omnivorous, and occupied a broader ecological range. Other theorists such as Fox (1991) have also cautioned against being "dazzled" by chimps because of our close phylogenetic relationship, noting that human ancestral apes splitting off from the hominid line had different organizational potentials due to their distinct forms of locomotion and ecological adaptations that extended well beyond the forested habitats of contemporary apes. In short, there is no hard evidence that the social behaviors of modern chimpanzees, such as male philopatry, territoriality, and aggressiveness, represent seminal and enduring hominin traits. The counter-argument is that parallel features of human and nonhuman primate social organization may be a reflection of selective factors common to the environmental conditions in which they currently live rather than a behavioral template that was genetically imprinted in the ancient past.

As referenced above, avatars for ancient male philopatry have recently been sought through the analysis of hominin fossil remains. In the case of the aforementioned study by Copeland and colleagues (2011), the finding that australopithecine teeth sex-typed as female showed isotopic profiles inconsistent with local geological substrates was taken to mean that females were more likely than males to have dispersed from their natal areas. The validity of such study conclusions, however, must be viewed with caution for at least two reasons. First, as the authors acknowledge, gender identification of these fossil remains is problematic due to the overlap in tooth size among male and female hominoids and, as in this case, to the absence of complete *in situ* skeletal substrates. Second, while the methods and findings of this study are intriguing, generalizations about landscape use by Pliocene taxa as diverse as australopithecines cannot be made with any confidence, particularly on the basis of such a limited sample.

Among the more interesting observations of the study is that the sexes may have exhibited distinct subsistence ranges during their lifetimes. In addition, the researchers question the notion that australopithecine social organization resembled that of contemporary apes since the relatively minor dimorphism in the size of canine teeth indicates a low degree of male competition. They note: "We think it is likely that there is no appropriate modern analogue for the social structure of these australopiths, given their marked anatomical and ecological differences from extant hominoids" (Copeland et al. 2011: 77).[6]

Finally, utilization of hamadryas baboons as a model for the evolution of hominin multilevel social organization is highly selective. The impression given is that male philopatric structure is somehow more "evolved" than its female counterpart. Greuter et al. (2012) acknowledge that multilevel or modular social organization occurs among other papionins, such as savanna and gelada baboons, in which female philopatry and matrilineal dominance hierarchies prevail. However, despite the close phylogenetic relationship of these papionins with hamadryas baboons, the authors make no attempt to explain how or why such contrastive multilevel systems evolved. Instead, savanna and gelada groups are seemingly disqualified as suitable models for the evolution of hominin social organization due to their deviation from the presumed *Pan-Homo* prototype. The proposed evolutionary pathway for hominin social systems is thus skewed toward a set of hypotheses that marry intraspecific threat and the perceived need for male defense and provisioning of females with the mitigating institutions of pair bonding and multilevel fraternal interest groups. As demonstrated by savanna and gelada baboons, however, complex organization among primates may arise on more than one type of social platform, the nature of which appears to vary with environmental conditions.

In summary, ancestral male kin hypotheses establish an evolutionary model that essentially freezes the critical interplay of biology and environment at a single point in time—namely, at the separation of our *Pan-Homo* ancestor from the shared hominid line. This divergence is seen as marking the genesis of an imprinted *biogram* that predisposed emerging hominins toward social patterns driven primarily by male reproductive strategies. In such models, distinguishing human trait complexes evolved as elaborations or cultural embellishments on this prototype, the deviations from which are viewed as aberrant or maladaptive.

Ecology and Female Choice

While ancestral male kin hypotheses seek to cast our ancient forebears into a single mold, ecological models approach the reconstruction of hominin social life by proceeding in the opposite direction. Namely, they prioritize the investigation of morphological and behavioral *diversity*, its basic parameters, and its probable antecedent conditions. When applied to human evolution, this method is more likely to generate multiple prototypes in time and space. Ecological models provide insights on how the distribution of resources in a given niche influence the spatial distribution of females and males, and how, in turn, these factors shape both the reproductive strategies and social fabric of ancient and modern primate groups.

Early explorations into the importance of ecology in the evolution of social behaviors shifted the focus of inquiry from unilateral male mating systems to female reproductive strategies. Orians (1969), for example, examined a spectrum of bird and mammal mating patterns in an effort to identify the common conditions favoring polygyny. Since, he reasoned, polygyny should always enhance the reproductive success of males, its presence or absence must rest on its relative benefits or costs to females. He concluded that female choice to enter into a "bigamous" union was influenced by the quality of habitats occupied by available consorts. The so-called *polygyny threshold* was crossed when females perceived that the opportunities for resources and parental care offered by mated males were greater than that available with unmated males. In other words, females choose polygynous relationships only when it enhances their reproductive success. Similar conclusions were reached in cross-species mating studies by Emlen and Oring (1977: 217).

Wittenberger (1980) argued that females in nonterritorial species need not pursue polygynous unions in order to access favorable habitats. Instead, he proposed that polygamy among social mammals was a result of selective factors favoring the formation of female social groups. "Sociality" was seen as maximizing the lifetime reproductive output of individual female group members by enhancing cooperative defense, the nurturance of offspring, subsistence efficiencies, and maintenance of close kin ties.

Two studies authored by Wrangham (1979, 1980) shed further light on the ecological underpinnings of primate social systems. These basic research findings are summarized in Table 2.2.

Wrangham's 1979 study focused on ape social patterns, and on male and female intrasexual competition as the principal driver of behavioral strategies for reproductive success. Wrangham observed that, in contrast to smaller primate species, the body size and metabolic rate of apes favor individual rather than group foraging by females to minimize food competition. Male reproductive strategies vary depending

Table 2.2. Impact of ecological variables on female feeding strategies and primate social organization.

Food Distribution	Territoriality	Female Feeding Strategy	Male Distribution	Intrasexual Relationships	Intersexual Relationships	Primate Example
Discrete, high-quality, high density food patches	Territorial; resource density allows defense of home range	Permanent female-bonded (FB) kin groups with social feeding	Polygynous one-male groups (OMGs) [Also assumes presence of autonomous bachelor all-male groups (AMGs)]	*Females*: Competitive ranked kin groups with differential access to high-quality foods; cooperative against outsiders. *Males*: Competitive dominance hierarchies for OM group leadership	*Females*: Capitalize on male competition by favoring one male over another. *Males*: Exchange protective and defensive roles for sexual access	Japanese monkey (*Macaca fuscata*)
Seasonally variable: High-quality patches (growth diet) vs. lower quality, uniformly distributed forage (subsistence diet)	Nonterritorial; day range incompatible with area that may be effectively defended as exclusive territory	Permanent FB kin groups with social feeding; may disperse in times of scarcity	Seek membership in multimale-multifemale groups (MM-MFGs) Bachelor AMGs	*Females*: Dominance hierarchies and feeding competition among ranked kin groups. *Males*: Competition for MM-MFG entry and sexual access	*Females*: Manipulate group size by number of males they accept. *Males*: Compete for group membership	Olive baboon (*Papio anubis*)

Subsistence diet with uniform distribution of foods; no high-quality food patches	Mutually exclusive territories	Females forage alone in ranges that they can defend	Exclusive territories shared with consorts in monogamous pairings	Competitive, with both sexes defending exclusive territory	Cooperative, with both sexes investing in pair-bonded offspring	Gibbon
Uniform food patches for both growth and subsistence diets *Or:* Subsistence diet distributed in small high-quality patches	*Females:* Dispersed, undefended, overlapping territories *Males:* Larger core ranges and defense of boundaries	Permanent male consortships and social feeding (gorilla) *Or:* Females forage alone, as food density distribution prevents females traveling together (chimpanzee, orangutan)	Bisexual groups with one dominant male (gorilla); *Or:* Solitary males (orangutan) *Or:* Multimale groups (MMGs) based on male philopatry (chimpanzee)	*Females:* Avoid food competition; may attack outsider females entering their core area *Males:* Competitive in OMGs and solitary life; cooperative in MMGs for defensive activity	*Females:* Alliances with dominant males in OMGs; mate with multiple males in MMGs to cement alliances; no ties with solitary males *Males:* Affiliative and peaceful relations with in-group females vs. outsiders	Great Apes

Source data: Wrangham (1979, 1980).

on female distribution. In general, females forage alone except where food sources are unusually plentiful. They may also join permanent groups where resources are abundant but dispersed, and of lower quality (as among gorillas). In both cases, Wrangham notes that a female ape's best option is to align herself with a dominant male, one who can most effectively protect her and her offspring from aggressive acts by other male conspecifics. Where male kin are localized, as among chimpanzees, females may pursue multiple sexual liaisons with members of male-bonded groups to further cement their support and defense alliances.

Female apes do not typically defend exclusive territories because their core foraging range is greater than their daily subsistence range. Monogamy therefore offers no advantage if a female cannot defend her boundaries. Where exceptions occur, as with lesser apes such as gibbons, females may optimize their reproductive success by entering into a pair-bond relationship with an unattached male willing to provide territorial defense. This pattern is exceptional among primate species, however, and only occurs where the foraging territories, interests, and inclusive fitness of both sexes coincide.

Wrangham's 1980 paper focused specifically on species with the most common primate social pattern, namely, that based on residential groups of related females. Species were categorized as female-bonded (FB) where females breed in their natal group, where there is an intergroup transfer of males, and/or where females engage in consistent relationships, including mutual grooming, aiding, huddling, and aggression. FB species were found to occur where high-quality food sources are distributed in discrete, defensible patches. In such niches, related females form alliances for mutual defense of feeding patches, arranging themselves into highly differentiated networks and competitive intrasexual relationships. Female dominance hierarchies often define differential access to quality feeding areas and may also impact the reproductive success of individual group members.

Wrangham argued that the relative density of these discrete feeding patches shapes both the optimal foraging strategies for females and the prevailing form of male competition. Where the density of patches is sufficiently high to allow for defense of a home territory, female social feeding within one-male groups predominates. In these instances, females take advantage of male sexual competition by favoring one male over another. Female choice of a dominant male aligns their interests for protection from the aggressive behavior of male subordinates with the consort's interest in keeping competitors away.

In contrast, where discrete defensible feeding patches are distributed over a wider geographical area, such as in more open habitats, FB social groups are nonterritorial in nature. Wrangham linked the seasonal distribution and variable quality of food resources with growth versus subsistence diets, the former associated with concentrated high-quality feeding patches, and the latter with lower-quality, uniformly distributed food sources. FB groups may aggregate and disperse in concert with cycles of resource availability. In such habitats, the distribution of essential food resources exceeds the range that may be effectively defended by a single male. Instead, permanent FB groups rely on the selective recruitment of multiple males to expand their group size, a primary factor in successful food competition with neighboring communities. Females manipulate group size by regulating the number of males they accept. Wrangham noted that adult male membership in FB groups is competitive. As outsiders, males are at a disadvantage since, unlike the members of highly structured female networks, they have no allies. Female choice is therefore expected to favor cooperative rather than aggressive males for group membership.

In summary, ecological studies have shed new light on primate social structure when viewed from the perspectives of both female and male inclusive fitness. Some general conclusions may be drawn. First, the basic primate social unit is the matricentric family, consisting of a female and her offspring. This kinship unit is often multigenerational and may be linked to others in a highly differentiated network of alliances that structure intra- and intergroup access to subsistence and reproductive resources. Second, because female inclusive fitness is linked to food, the nature and distribution of food resources and associated optimal feeding strategies are the primary determinants of female sociospatial distribution. Third, because male inclusive fitness is linked to access to fertile females, the nature of male competition is shaped by female distribution. According to Wrangham:

> If FB groups have evolved as modelled, they support the view that male strategies are ultimately the result of female distribution; *i.e.* males compete for access to given clumps of females in a system of "female defense polygyny" ... Consequently the number of females per group and the number of males per female are considered to depend ultimately on the strategies of females, rather than on the ability of males to dominate each other in competition for females, as is sometimes implied. (1980: 287–88)

If the nature of ecological niches and the distribution of resources therein shapes the optimal feeding strategies of female primates, then the resulting social systems may be predictable in general outline.

Female Foraging and the Matricentric Family

Recognition of the matricentric family as the basic primate socioeconomic unit has significant implications for hominin evolutionary theory. If ancient first families consisted not of one-male units (OMUs) or pair-bonded couples, but of mothers, their offspring, and close female kin; if joint subsistence activity, food sharing, and other cooperative relationships originated within this unit; and if the female-bonded group provided an enabling structure for the increasingly lengthy development and enculturation of offspring, then prevailing notions about natural, kin, and sexual selection among ancient hominins are cast in a new light.

Such was the basic premise of Nancy Tanner's insightful book, *On Becoming Human* (1981). Tanner presented a new model of human origins that assigns a central role to female gathering and to the matricentric unit as the cradle of cognitive, social, and technological advancements marking the hominin transition. This subsistence model, which will be considered in more detail in chapter 3, represents a significant departure from traditional evolutionary theory in three important ways. First, consistent with the academic awakenings of the 1980s, it recognizes that natural selection has operated on not one, but both sexes. It proposes that early female hominins increased their reproductive success, and hence their inclusive fitness, by honing skills for the effective provisioning of offspring and for the transmission of knowledge essential to the survival of their issue to maturity. Second, it recognizes the female-bonded social group as the logical unit in which kin selection occurred in ancient times. Tanner observed:

> In the context of this long term mother-offspring and sibling association, ties among children of the same mother developed. These continuing mother-offspring and sibling relationships made kin selection possible. Mothers and offspring share many of the genes that are ordinarily in free variation in a population, and siblings with different fathers—as would ordinarily be the case—also share some of these genes, although not as many as mothers and their young. Therefore, behavior among kin that favors mutual survival serves to pass on shared genes. (1981: 162–63)

These same matrilaterally extended kin ties are seen as providing the glue for cooperative activities, such as food sharing and mutual care of dependent young. Maternal lines in which such relationships were developed had greater reproductive success and hence were favored by kin selection. Tanner thus broadens the concept of "parental investment" in offspring to include the collective investment of mutual kin, rather than the variable contributions of sexual partners. Said Tanner of early hominins:

At this stage males had very little direct investment in their offspring. Females had the biological investment of internal gestation, birth, and nursing and were providing most of the physical care *and* food, as well as much of the protection for their infants. Although some food and protection were supplied by males, sexual partners were only occasionally and irregularly involved; brothers and sons who frequently associated with mothers and sisters probably were the males who most regularly contributed animal protein and protection. (1981: 163)

A third point of departure from traditional models involves the role of sexual selection in early human society. Male-centered models typically see sexual selection as an adjunct of male-male competition, where males vie to attract mates through the possession and display of exceptional characteristics or behaviors. Thus Fox (1972) posed the theory, apparently concurred in by Wilson (1975), that sexual selection among early hominins was linked to male prowess in hunting, tool making, leadership skills, and other attributes seen as central to male provisioning and group survival. As in the case of Dawkins's (1976) "he-man" theory of female choice, selective pressures are seen as favoring male behaviors that *differentiate* them from females and that reflect the proposed division of labor between provider/defender and domestic/nurturance roles.

Tanner, in contrast, joined Wrangham and others in focusing on the implications of female choice for reproductive success within the framework of a pre-existing matricentric socioeconomic unit. When viewed in this context, females select sexual partners who most effectively insulate and protect them from male aggression and who exhibit sociable, affiliative, and cooperative behaviors. From an evolutionary perspective, female selection of "friendly" males is seen as playing a central role in the divergence of hominins from ancestral apes:

> What may have been selected for among the transitional hominin males was the capacity to be extremely social but yet sufficiently aggressive when required and an ability to make fine discriminations as to situational necessity. Thus, the males of the transitional population would come to more closely resemble the females than had the males of the ancestral population. (Tanner 1981: 165)

Tanner proposes that the combination of female parental investment and female choice of sociable, intelligent consorts had an impact on gene frequencies over time—that natural selection and sexual selection worked hand in hand to, in essence, "domesticate" males and thereby shape the character of early hominin society. Recent neurobiological evidence that suggests a genetic basis for this premise is explored more fully in the discussions of mosaic brain evolution in chapters 5 and 6.

Cooperative Breeding and Eusociality

Sarah Hrdy, in her comprehensive works *Mother Nature* (1999) and *Mothers and Others* (2009), examines the evolutionary significance of motherhood, maternal investment, and female choice within the broader context of cooperative breeding communities. Cooperative breeding, simply defined, refers to the shared care and provisioning of young. Although relatively uncommon in the animal world, this reproductive pattern occurs across a diverse range of invertebrate, bird, and mammalian species, including humans (Hrdy 2009: 177). Shared care, or *alloparenting*, relies on the evolution of altruistic behaviors among members of a breeding community.

Offspring among cooperative breeders benefit from the presence of multiple caregivers and providers, allowing a longer period of growth and development while lessening the burden on their mothers. Hrdy argues that the successful rearing of offspring among early hominins required the efforts of several caretakers or alloparents, likely within the context of a multigenerational extended family. In Hrdy's view, cooperative breeding provided the pre-existing condition for development of cognitive and emotional traits that distinguish hominins from both ancient and modern apes:

> I don't think humans ended up with greater inter-individual tolerance, aptitudes for mind reading and learning, and with them greater capacities for cooperation than other apes because they already possessed sapient-sized brains, symbolic thinking, and sophisticated language. Rather, I am convinced that our line of hominins ended up with these attributes because of an *unprecedented convergence*—the evolution of cooperative breeding in a primate already possessing the cognitive capacities, Machiavellian intelligence, and incipient "theory of mind" typical of all Great Apes. The ancestors of humans started from a difference place than chimpanzees did. (Hrdy 2009: 279–80, emphasis added)

While alloparental care is not necessarily limited to close kin, natal kin networks provide ready opportunities and incentives for food sharing, mutual defense of heritable resources, and other group activities. Hrdy notes that the greatest benefit of philopatry for females is the presence of female kin. As among other long-lived mammals, matrilineal kin not only provide for the mutual care of offspring but are the fountainhead of critical knowledge about local food resources that is passed on to the next generation.

There are only three species in which females live beyond reproductive age—elephants, pilot whales, and humans (Hrdy 1999: 273). In each case, postmenopausal females compose a kind of "sterile class" capable of contributing food and care to needy kin, as well as vigorously defending weaned offspring from the aggressive acts of conspecifics and other predators. The "grandmother hypothesis" supports the view that selective pressures favored the evolution of long lifespans for postmenopausal females due to the positive contribution they provide to the nurturance and survival of their grandchildren (Hawkes, et al. 1998; Hawkes and Blurton Jones 2005).

A parallel concept to cooperative breeding is *eusociality*.[7] Eusociality is defined by Wilson (2012: 133) as ". . . the condition of multiple generations organized into groups by means of an altruistic division of labor." Eusocial species vary in the degree to which altruistic acts and reproductive roles are strictly defined. At one extreme are eusocial bee colonies, where helpers devote a lifetime to the provisioning and protection of a queen's offspring, while forgoing their own reproduction as a sterile caste. At the other end of the continuum are hominins and other mammals, where all group members are capable of reproducing and where flexible alliances and collaborative activities impact both individual inclusive fitness and group success.

According to Wilson, ancestral humans evolved as a eusocial species through selective processes at the individual and group levels. While recognizing that inclusive fitness has operated to favor altruism among multigenerational members who share a close genetic relationship, Wilson argues that kin selection alone does not explain the evolutionary processes essential to the formation of early human social groups. Instead, he proposes that all species with evolved eusociality began with the concentration of their members at protected campsites at which a division of labor for reproductive success emerged. Namely, some group members remained at the "nest" to defend and nurture dependent young while others ventured away from the encampment in search of food to be shared. Competition and the maintenance of complex alliances within the group favored the refinement of "social intelligence." At the same time, altruistic behaviors that elevated interests of the collective over that of individual members gave the group a competitive edge vis-à-vis outsiders for territorial dominion and access to resources.[8] Says Wilson:

> . . . . we can expect that the outcome of between-group competition is determined largely by the details of social behavior within each group in turn. These traits are the size and tightness of the group, and the quality of communication and division of labor among its members. Such traits are heritable to some degree; in other words, variation in them is due in part to differences in genes among the members of the group, hence also among the groups themselves. The genetic fitness of each member,

the number of reproducing descendants it leaves, is determined by the cost exacted and benefit gained from its membership in the group. These include the favor or disfavor it earns from other group members on the basis of its behavior. The currency of favor is paid by direct reciprocity and indirect reciprocity, the latter in the form of reputation and trust. How well a group performs depends on how well its members work together, regardless of the degree by which each is individually favored or disfavored within the group. (2012: 53–54)

Wilson describes the resultant human genetic code as a "chimera," with one part prescribing traits that favor individual reproductive success at the expense of other group members, and one part prescribing altruistic and cooperative traits that advance the interests of the group in competition with others. The eusociality model thus provides a way of addressing not only the evolution of distinctively human traits but a process by which niche partitioning and speciation may have occurred in ancient times.

Anisogamy and Inclusive Fitness Revisited

The evolutionary significance of anisogamy is that it prescribes distinct but complementary reproductive strategies for females and males. Traditional androcentric theories assume that male strategies have predominated from ancient times, largely at the expense of females, through parasitism, coercion, and the sexual contract. More recent investigations have focused instead on the parameters of female inclusive fitness. Female breeding choices, alliance networks, provisioning, and alloparenting are now recognized as pivotal factors in the architecture of both ancient and contemporary primate groups. Moreover, female reproductive strategies have been observed to vary in characteristic ways with the nature and distribution of resources in a given ecological niche, shaping, in turn, male intra- and intersexual relationships.

In light of these findings, an alternative model of hominin first families presents itself. The cornerstone of this model is necessarily the matricentric unit, either alone or linked with other such units in a social group that includes one or more mature males. Whether resultant social groups are primarily monogamous, polygynous, or polygynandrous is a consequence of the optimal feeding strategies of their female members.

Current data suggest that where subsistence-level food is distributed in uniform patches, as among gorillas, females and their young either feed alone or with other nonbonded matricentric units in one-male groups. The optimal feeding strategy for females is defined by their ability to competitively secure food resources for themselves and their offspring while enjoying a margin of safety from male aggression. Where resources are distributed within a defensible feeding range, as among gibbons, females may form a monogamous pair bond with an unattached male to help secure their territorial boundaries. More typically, however, the daily foraging distance is too short to allow for defense of an exclusive territory. In such niches, the common primate pattern is for individual matricentric units to attach themselves to a shared consort whose position of dominance in the male hierarchy secures the subsistence perimeter and the safety of group members.

In mosaic habitats, where food is commonly distributed in discrete, defensible high-quality patches, female feeding strategies and resultant social patterns are distinct. Matricentric units often congregate into permanent, ranked female-bonded kinship groups that structure access to both food and reproductive resources, often on a multigenerational basis. Social feeding among related matri-kin prevails. In these settings, females compete with one another for high-quality foods within the group's range, but cooperate to dominate their neighbors. They play an active role in directing the timing and movements of the group and may take part in aggressive intergroup relations. Where the density of food patches is sufficient to support group members and allow exclusive claim to a home territory, female-bonded kinship units may attach themselves to a single male who is up to the task of defending their boundaries. Where food patches are less densely distributed or more seasonally variable, female-bonded kin select multiple males who compete for group membership and who assist in maintaining the group's socioeconomic interests.

The influence of ecological factors on female feeding strategies, mating patterns, and ultimately on the structure of primate groups has significance for the reconstruction of early human social life. Paleoanthropologists generally agree that the hominin transition occurred during the Late Pliocene among selected bipedal ape populations occupying eastern and southern Africa. The precise ecology of this vast region between 5 and 7 ma is unknown. What *is* known, however, is that this period was marked by increasing aridity and the gradual displacement of forests by more open mosaic habitats. The prevailing theory is that proto-human apes transitioned over time from a primarily arboreal to a fully terrestrial existence in response to the effects of climate change.

Expansion into new habitats would have, by definition, involved a change in the nature and distribution of food resources. Mosaic settings, such as adjacent riverine forests, marshlands, waterways, lakes, and edaphic grasslands, likely offered patchy, high-quality foods, the distribution and availability of which may have waxed and waned with the seasons. Corresponding shifts in female feeding strategies and social patterns would have necessarily followed. Based on current ecological models, the probable direction of this shift would have been from autonomous matricentric units clustered into one-male groups to larger aggregates of female-bonded kin groups accompanied by multiple males. In other words, we would expect evolving hominins to be moving from individual to social foraging, from polygyny to polygynandry, and from loosely structured to more complex intra- and intergroup relationships.

Such economic and social transitions set the stage for hominin advancement. As noted, the local clustering of multigenerational female kin is perhaps the most logical social grouping for the genesis of alloparenting and food sharing. These developments likely appeared early on as a prelude to hominin encephalization. As Hrdy (2009: 176) quips: "Brains require care more than care requires brains." Similarly, as evidenced by other long-lived species such as elephants, alloparenting and joint provisioning may evolve independently within the social web of cooperating female kin, and in the total absence of male parental investment. For females, the combination of alloparenting and food sharing within the extended matricentric family allowed for increased infant dependency, prolonged mother-child relationships, enhanced opportunities for the cognitive development of young, and shorter birth intervals—all giant steps for offspring survival and female reproductive success.

Changes in female socioeconomic strategies accompanying the terrestrial adaptation would have also set in motion significant changes affecting male inclusive fitness. The shift from one-male groups to multimale groups not only enhanced the number of males with access to fertile mates, but modified, in part, the basis on which such access was attained.

In the case of polygynous groups, male intraspecific competition set the bar for access, with the dominant male able to attract an entourage of several females and their offspring. Such males served, in essence, as "sentries with benefits," defending the group's perimeter from outside aggression while siring the majority—although not all—of the group's offspring during their tenure. Significantly, the majority of males, who leave their natal group prior to sexual maturity, may have limited opportunities to reproduce during their lifetimes. In polygynous social systems, male inclusive fitness relies primarily on movement up the rungs of the dominance hierarchy, or, to a lesser extent, on the occasional ability to woo receptive females into a clandestine coupling.⁹ Such systems, by definition, limit the genetic diversity of offspring and select for sexually dimorphic traits.

In contrast, it is likely that the clusters of female-bonded social groups characteristic of new terrestrial adaptations incorporated multiple males in polygynandrous mating arrangements. Male dominance hierarchies would have continued to be a factor in sexual access, but female choice would play an increasingly influential role in determining the acceptance of individual males into group membership. As suggested by Tanner and Hrdy, it is probable that males incorporated into early hominin social groups were selected by females for behavioral traits that enhanced the protection and survival of their offspring.

While this bias in female choice is observable among some contemporary primates, its effect in ancient times would have been accelerated if alloparenting and food sharing were already established in the proto-human domestic unit. In other words, mature males seeking to join a female-bonded social group may themselves have been raised in a cooperative food-gathering and food-sharing environment prior to leaving their natal group. Cementing a consort relationship with similar learned behaviors may have had less to do with the paternity of a female's existing offspring than a male's prospect of future progeny. As Hrdy has observed: ". . . when a male primate's reproductive success is substantially enhanced by assisting his mate rear offspring, and when he has no better reproductive alternatives . . . he helps" (1999: 213); and again: "The risk to a male's posterity from caring for another male's offspring is outweighed by the still graver risk of dying childless" (2009: 89).

Female sexual selection of friendly, helpful consorts would have an effect not only on male reproductive success but, over time, on gene frequencies as well. Sexually dimorphic traits among primates that provide an advantage in male-male competition, such as body and canine size, are commonly associated with polygyny (Clutton-Brock and Harvey 1976). With the shift toward polygyandrous mating systems, a gradual reduction in sexual dimorphism would be expected. Tanner observed:

> If females were choosing to copulate with the sociable males—ones who did not bare large canine "fighting teeth" at them—then sexual selection supported reduction of male canines. This could reinforce the concurrent natural selection for reduced canines (for a more effective chewing apparatus) that was acting on both sexes. From the fossil record, canine reduction appears to have occurred very rapidly indeed. (1981: 271–72)

In summary, the movement of proto-human apes from forest niches to contiguous mosaic and savanna habitats catalyzed new socioeconomic adaptations that impacted the reproductive success of both sexes. At the individual level, female inclusive fitness was enhanced by the development of alloparenting and food sharing within female-bonded kinship units. Male inclusive fitness was similarly enhanced by greater access to fertile females and by an increasing investment in the group's offspring. Intrasexual competition would be expected to continue to structure female access to high-quality resources as well as male access to females, but to be counterbalanced by a trend toward affiliative intersexual relationships. These cooperative relationships may have been founded on the enlistment and joint participation of males in gathering activities and the rearing of young.

Multilevel Selection Revisited

At the same time that natural selection was targeting traits at the individual level, the evolutionary process was also operative at the group level. Laying claim to foods distributed in discrete, defensible patches required early hominin community members to engage in collaborative efforts that advanced the common good in relation to neighboring groups, i.e., to secure sufficient fitness-related resources to enhance the reproductive success of group members. Wilson's theory of multilevel, eusocial evolution proposes that selection would favor those groups with the greatest degree of internal cooperation and cohesion—social strategies that are viewed as at least partially heritable. Wilson goes on to argue that group formation, or tribalism, along with intergroup conflict, is an ancient and fundamental human trait:

> Our bloody nature, it can now be argued in the context of modern biology, is ingrained because group-versus-group was a principal driving force that made us what we are. In prehistory, group selection lifted the hominids that became territorial carnivores to heights of solidarity, to genius, to enterprise. And to *fear*. Each tribe knew with justification that if it was not armed and ready, its very existence was imperiled. (2012: 62)

Similar notions about the aggressive, territorial nature of early human groups have been presented by Ardrey (1966), Tiger (1969), Tiger and Fox (1971), and, more recently, Wrangham and Peterson (1996). While there is considerable evidence to support the proposition that intergroup conflict is linked to resource competition, there is less of an empirical basis for extrapolating late Pleistocene or Neolithic scenarios of intertribal warfare back to the late Pliocene. Perhaps the relevant question in reconstructing ancient social life is the extent to which essential resources were scarce or contested in relation to population, and ways in which early hominins may have adapted their social strategies to fit the carrying capacity of their primary feeding ranges.

Contemporary primate female-bonded species occupying niches with discrete, defensible food patches are nonterritorial in nature, except in those instances where the density of these patches makes daily enforcement of the resource perimeter feasible. In the more open mosaic habitats, which are arguably most comparable to those occupied by transitional hominins, relationships among neighboring groups are generally stable, with dominance based on group size. Small groups yield to larger ones in intergroup encounters. Group size is, in turn, regulated by females by manipulating the number of males admitted for membership. Group size may also be adjusted in response to seasonal changes in resource types and availability, with permanent kin groups dispersing into subgroups in times of scarcity. This manipulation of group size by females optimizes foraging strategies for both growth and subsistence diets while maintaining stable intergroup relations.

Returning to the question of political relations in the Late Pliocene, there seems no reason to propose intergroup conflict or aggression as a necessary driver in hominin evolution. How, in lieu of conflict and open hostilities, could one population exert dominance over another? By simply producing greater numbers of fertile offspring over time. As succinctly stated by Hrdy: ". . . from an evolutionary perspective, child survival was the currency that mattered" (2009: 105). It mattered not only for individual fitness, but for the adaptive advantage realized by breeding groups that had mastered the skills of intra- and intersexual cooperation. It is proposed here that selective factors in hominin evolution favored social organizations that successfully marshaled the metabolic and genetic contributions of *both* sexes for the perpetuation of their members. In other words, selection favored structures and internal social networks that allowed group members to not only outsmart but *outbreed* their competitors.

An analogy akin to energy capture is useful here. For primate communities, the ecological niche they occupy and share with their neighbors contains finite resources that must be acquired, metabolized, and converted into energy for sustenance and reproduction. As a natural consequence of anisogamy, the metabolic contribution of the sexes to the survival of offspring differs, with females assuming the primary burden of gestation, birth, lactation, and early childhood nurturance. The ratio of male and female metabolic contributions to the next generation, as well as their genetic contributions, has been observed to vary with the nature and distribution of food sources.

For example, in polygynous groups, mother-child units are essentially on their own metabolically, in competition not only with other females, but with the bulk of males as well. Males consume their share of resources in the same niche, but are *noncontributors* in the provisioning of young. A parallel situation is reported by Dublin (1983: 305) for African elephants, where young bulls may be the primary competitors of matri-clan heads and their offspring for food, water, and mineral deposits. As among many primate groups, the female dominance hierarchy serves to regulate access of group members to both food and reproductive opportunities. Dominant matriarchs may effectively limit the fecundity of subordinates, thereby manipulating group size in relation to available resources.

Where polygyny prevails, the majority of males may be noncontributors to the gene pool as well. Indeed, depending on the tenure of dominant males, a considerable percentage of a group's male membership may eat, but fail to mate or to otherwise make a substantive contribution to the next generation. Such breeding populations are also less diverse genetically, and hence less resilient and responsive to environmental change.

Successful colonization of mosaic habitats by proto-human apes and the new socioeconomic adaptations this required represented the vanguard of more productive forms of energy capture and reproductive success. Cooperative breeding communities had their genesis in female-bonded matricentric families that adopted alloparenting and food sharing strategies. The initial division of labor in early hominin groups, therefore, was likely by age rather than sex. Alloparenting made possible the marshaling of metabolic energy from two classes of genetic noncontributors, namely, premenstrual and postmenopausal females, for the care of young. This labor division freed mothers for more intensive foraging activities, at least some of which may have been undertaken cooperatively with other group members.

When viewed from this perspective, multimale-multifemale groups offered certain advantages. The integration of selected males into female-bonded social groups that were already engaged in cooperative breeding added a new dimension to labor division. Male participation in food gathering and food sharing may have arisen as an extension of mating relationships and of female choice. The addition of cooperative males enhanced a group's survival by turning food competitors into food sharers and by helping to establish dominion over their neighbors through a swelling of their ranks. Groups with multiple males also enjoyed the benefits of polygynandrous mating opportunities and greater genetic diversity over time.

Natural selection may thus have favored breeding populations whose members were organized in ways that most successfully captured the metabolic and genetic resources of both sexes. By converting male noncontributors into contributors and by maximizing labor division by age and sex for the collective benefit of current and future members, eusocial groups were able to increase their numbers and prevail over others. Rather than being dispersed into small nomadic bands, the most successful proto-human communities in the Late Pliocene were perhaps the *largest*, composed of several permanent kinship groups with heritable and defensible rights to premier sources of water and high-quality food patches.[10]

Solving Galton's Problem

This chapter has presented divergent perspectives on our *Pan-Homo* origins. Efforts to reconstruct an ancestral ape prototype have typically relied on two primary sources of data: (1) observations of contemporary nonhuman primates and (2) cross-cultural studies of contemporary human societies. The objective of such investigations is to identify common interspecific behavioral traits that may be indicative of our shared genetic heritage and hence provide a window into our primate past. The architects of social origins theories have faced two methodological challenges: (1) the representativeness of contemporary human and nonhuman primate data for defining universal traits, and (2) the relevance of these synchronic data sets for the creation of diachronic models.

In the late nineteenth century, anthropologist Sir Edward Tylor (1889) conducted a now-famous cross-cultural study that linked changes in unilineal kinship systems with levels of societal complexity. His conclusions were challenged by Sir Francis Galton who argued, in essence, that inferences on evolutionary development derived from extant cultural similarities were invalid unless one controlled for extraneous variables, such as the effects of cultural diffusion. What became known as "Galton's Problem" has influenced the experimental design of modern comparative studies, including the development of standard cross-cultural samples that minimize the unintended effects of "autocorrelation." Such methodological refinements provide greater assurance that spurious correlations related to geographical proximity, linguistic affiliation, political complexity, or other externalities have been minimized by ensuring that a historical approach is taken in sample selection.[11] The translation of statistical significance into phylogenetic significance, however, remains problematic.

The *Ethnographic Atlas* is a large cross-cultural inventory compiled by G. P. Murdock in 1967. It codes a range of cultural characteristics for 857 societies worldwide and has been utilized by some researchers to identify patterns of social behavior that enjoy wide distribution and therefore might represent hominin universals. For example, data on postmarital residence patterns for this large sample, when taken in aggregate, have been interpreted to mean that male-centered groups predominate in almost 70 percent of human societies (van den Berghe 1979: 111). Notably, however, such gross tabulations fail to ensure that the raw inventory is itself representative, i.e., that measures have been taken to minimize the effects of autocorrelation attributable to factors such as historic relationships or geographical proximity. Is the observed skewing toward patriliny and patrilocal/virilocal preferences an indelible marker of an ancient biogram, or are prevailing patterns simply a byproduct of the global expansion of post-Neolithic and postcolonial political economies in recent times? Can the ethnographic present be presumed to represent the ethnographic past? When it comes to viewing the *Pan-Homo* ancestor through today's mirror, Galton's Problem remains.

Similar questions may be posed regarding the extent to which contemporary primates can be taken as representative of ancestral Pliocene apes. As suggested by primatologist Frans de Waal (2013), if phylogenetic closeness is the primary criterion for crafting ancestral prototypes, one could with equal alacrity point to bonobos as the more logical living descendants of *Ardipithecus ramidus*—to their peace-loving, altruistic, and female-dominant behaviors as the social foundation of subsequent hominin lineages:

> What if we descend not from a blustering chimp-like ancestor but from a gentle, empathic bonobo-like ape? The bonobo's body proportions—its long legs and narrow shoulders—seem to perfectly fit the descriptions of Ardi, as do its relatively small canines. Why was the bonobo overlooked? What if the chimpanzee, instead of being an ancestral prototype, is in fact a violent outlier in an otherwise relatively peaceful lineage? Ardi is telling us something, and there may exist little agreement about what she is saying, but I hear a refreshing halt to the drums of war that have accompanied all previous scenarios. (2013: 61)

Similarly, the fact that male philopatry predominates among contemporary chimpanzees and bonobos cannot be taken as proof that the social lives of ancestral Pliocene apes were organized on this same plan. One of the many insights provided by ethnographic and ethological field studies is that primate social behaviors are highly plastic. Female-centered and male-centered social groups are now understood to be importantly linked to reproductive strategies in specific environmental settings. These settings are not static, but have varied in time and space. Elimination of Galton's Problem would therefore seem to rest on a clearer definition of the ecological niches in which early hominins emerged and lived and the processes and conditions under which disparate reproductive strategies evolved.

Unfortunately, there are no surviving bipedal, terrestrial, brainy, and fully omnivorous nonhuman apes on which to base a model of our last common ancestor. Available evidence would suggest, however, that this Pliocene primate differed fundamentally in its ecological range, morphology, diet, life history, reproductive behaviors, and social life from modern chimpanzees. Similarly, there is an inadequate record of the full range of ancient human adaptations that were pursued in pristine Pleistocene settings. While historic Stone Age peoples such as the African Bushmen may provide a glimpse of what Paleolithic life was like in resource-poor habitats, they tell us little about how the occupation of richer, more biodiverse niches may have impacted the subsistence patterns and the size and structure of early human social groups. In short, constructing a de facto model of primeval hominin society is not as simple as drawing a straight line, however stair-stepped, between highly specialized forest-dwelling apes and highly marginalized hunter-gatherer populations.

Theories on the structure of ancient society at the time of the *Pan-Homo* split generally agree that the earliest hominin communities consisted of multimale-multifemale groups with promiscuous mating patterns. What happened from that point forward is the focus of debate—whether early hominids resembled aggressive chimps or more peaceful androgynous primates; whether dispersal at adolescence involved males or females; whether mating relationships were exclusive or nonexclusive; whether labor division and food sharing originated in the pair bond or in allocare networks; whether females or males were the primary providers; or whether intra- and intergroup relations were agonistic or collaborative.

Answers to these and other questions provided by current theories rely on disparate assumptions about Plio-Pleistocene ecology and its impact on hominin resources, subsistence, settlement patterns, and social demography. They also rest on an author's calculus of male and female natures, the prime movers of hominin life history changes, and the role played by mosaic brain evolution in shaping reproductive strategies and the architecture of human social groups.

Phylogenetic models help to define our biological relationships with other primates and the array of common genetic propensities that frame reproductive success, cognition, social technologies, and life histories. They highlight the epigenetic rules—the social DNA—by which gene-culture evolution proceeds. Ecological models, on the other hand, identify the selective factors that shape the phenotypic *expression* of those rules among both human and nonhuman primates in characteristic ways. Both of these perspectives are necessary to piece together the complicated puzzle of human origins. If there *is* such a thing as a genus *Homo* biogram, it is not a unitary, hardwired social template crafted in the ancient past, but rather the potential for behavioral plasticity—the ability to apply flexible strategies to meet the challenges of changing environmental conditions.

By laying claim to the best resources, and by developing collaborative social structures, some Late Pliocene lineages of advanced proto-humans succeeded at the expense of others. All of the core individual and group social traits that sustained early hominin taxa for millennia could have well evolved in lieu of pair bonding, big brains, territorial aggression, or systematic carnivory. But subsequent developments in the hominin line appear to be linked to the adoption of revolutionary subsistence and reproductive strategies—changes that catalyzed the distinctive morphological, life history, cognitive, and technological advances that define the genus *Homo*.

 3

Paleoecology and Emergence of Genus *Homo*

Where, when, and how some ancient hominin lineages crossed the murky threshold to humanity is arguably one of the premier questions for paleoanthropology. A number of solutions to this riddle have been proposed in recent decades, only to be challenged or undone by new additions to the fossil record and by a more complete understanding of Plio-Pleistocene climate effects. As our knowledge base expands, answers to questions surrounding human origins will continue to be a moving target.

Currently, the prevailing view is that a diversity of anthropoid apes occupied the African continent during the mid to Late Pliocene. At least some of these species were bipedal, were nominally omnivorous, and utilized crude stone tools. The general consensus is that the first humans evolved from this phylogenetic base around 2 ma in the eastern or northeastern regions of the continent.[1] Emergence of the genus *Homo* occurred during a period of marked climatic instability and is thought to have been set in motion by the challenges that proto-human apes faced in adapting to dynamic landscapes and habitats. Human origins are commonly associated with the appearance of *Homo erectus* (and some pre-*erectus* forms such as *Homo habilis*), a new clade of hominins that were bigger-bodied, bigger-brained, omnivorous, socially advanced, and more far-ranging than earlier Pliocene apes. *H. erectus* lineages demonstrated the unprecedented ability to adapt to a variety of biomes, successfully radiating throughout the tropical and middle latitude belts of Africa and Eurasia over the ensuing millennia. The primary catalyst for this giant step in human evolution is the subject of continuing scientific inquiry and debate.

This chapter will examine a range of theories on the socioeconomic life of early humans in Late Pliocene Africa. These theories generally agree that changing climate conditions played a central role in shaping the emergence and subsequent evolution of the genus. They differ, however, in two important respects. First, they make contrastive assumptions about the prevailing ecological setting in which early humans lived. As noted in the previous chapter, the nature of this setting has important implications for resource types and distributions, as well as the optimal socioeconomic strategies that may have been pursued for their exploitation. Second, there is fundamental disagreement about whether such ecological settings and the characteristic human adaptations with which they are associated were relatively uniform in time and space. Theories relying on a single prototype have historically proposed that distinguishing human traits evolved as a bundle in direct response to survival challenges in a specific type of habitat. Other theories emphasize the heterogeneous nature of Pliocene microenvironments and the essential plasticity required by early *Homo* to calibrate their socioeconomic strategies with changing conditions. Such models see Plio-Pleistocene hominins as polymorphic species that most likely displayed a variety of adaptive morphological, social, and behavioral traits.

Romance of the Hunt

The morphological characteristics that distinguish *Homo* from earlier hominins, such as significant increases in body and brain size, have historically been pointed to as evidence of increased dietary breadth, along with the potential for cooperative subsistence endeavors. The new dedication of *Homo* to omnivory, and in particular to meat eating, fit nicely into the portrait of males as cooperative hunters and providers. Reliance on hunting, when coupled with accumulating data on dramatic vegetation changes in post-Miocene Africa, aligned such dietary changes with the expansion of hominins from forest homelands onto the open grassy plains of eastern and southern Africa. It was in this arid savanna setting and under these special adaptive circumstances, some propose, that all the uniquely human traits evolved. This monotypic view has been perhaps the most persistent theme in human origins theory.

Academic interest in this hypothesis was launched in 1924 with Raymond Dart's discovery of a 2.5-million-year-old juvenile *Australopithecus africanus* skull (the Taung child) in South Africa. This find led Dart (1925) and, later, Robert Broom (1933) to propose that early bipedal proto-humans emerged from the forest and onto the arid savanna grasslands as hunters—aggressive carnivores who needed meat for survival. Dart (1953) later advanced the "killer ape hypothesis"—the idea that such aggressive tendencies are an instinctive and enduring human trait. A similar theme was elaborated on by Konrad Lorenz (1966) and popularized in a trilogy of works by Robert Ardrey (1961, 1966, 1976).

What became known as the "hunting hypothesis" was also born out of influential works emanating from two international academic conferences, the Social Life of Early Man in 1959 and Man the Hunter in 1966.[2] The resultant body of theory on human social origins was founded on two basic premises that largely mirrored Darwin's thoughts in *Descent of Man* nearly a century before: (1) that this transition was a consequence of arboreal abandonment by proto-human apes and their subsequent occupation of savanna grasslands, and (2) that cooperative hunting by males and dietary reliance on meat provided the necessary catalyst for human advancement. Architects of the hunting hypothesis formulated their basic assumptions on the paleontological and archaeological evidence available at the time, embellished with cultural musings on the implications of these data for ancient social life.

The initial premise, namely, that open grasslands made up the critical habitat in which early hominins evolved, is also referred to as the "savanna hypothesis." By the 1960s, it was well known that Early Pleistocene glacial advances in the northern latitudes of Eurasia triggered dramatic climate change effects on the African continent. A macro trend of increasing aridity led, over time, to desertification, along with the depletion of tropical forests and their displacement by grasslands. Once on the open plains, it is proposed that proto-human apes were subjected to new selective pressures that favored bipedalism and the shift to dietary reliance on meat eating.

The second basic premise of the hunting hypothesis is that hunting by males is the primary driver for development of the suite of traits thought to define the human genus, including bipedalism, tool making, encephalization, intelligence, modern infracranial morphology, the invention of fire, and symbolic communication. Once set in motion by cooperative hunting activity, these traits were seen as evolving rapidly as a package in a kind of domino effect, a process referred to by Wilson (1975: 68) as "autocatalysis." In short, hunting's significance for the evolution of all things human is global in scope. As expressed by one advocate:

> Hunting is the *master behavior pattern of the human species*. It is the organizing activity that integrated the morphological, physiological, genetic, and intellectual aspects of the individual human organisms and of the population who compose our single species. *Hunting is a way of life, not simply a "subsistence technique"* which importantly involves commitments, correlates, and consequences spanning the entire biobehavioral continuum of the individual and of the entire species of which he is a member. (Laughlin 1968: 304, emphasis added)

As discussed in chapter 2, hunting is also viewed by some as laying the cornerstone of human social groups, giving rise to both male-female pair bonding and the division of labor by sex. Isaac (1978) proposed that early hominin social organization was based on the establishment of central foraging points or "home bases" from which providers would wander and then return with high-quality foods. Male provisioning of females and young with meat at "protected nest" sites is also a central ingredient in Wilson's concept of eusociality (2012: 41–44).

Hunting activities have also been credited with a lengthening of the lower limbs in *H. erectus,* which some observers estimated to have reduced energy requirements by half and to account for a twofold increase in home range as compared to australopithecines (Anton, Leonard, and Robertson 2002; Steudel-Numbers 2006). Increases in *Homo* relative hind limb length have also been linked to enhanced running abilities associated with the pursuit of prey animals. These theories, however, have found little support as the number of fossil specimens available for comparison has grown. It now appears that australopithecines, *H. habilis,* and *H. erectus* all have limb lengths within modern ranges, and hence had similar bipedal capabilities (Pontzer 2012).

Recent paleontological evidence has added new wrinkles to hunting hypothesis assumptions about the way that human traits evolved. Hominin fossil discoveries in the past two decades suggest that the course of human evolution, like politics and sausage making, is a messy business. It is now clear that reconstruction of our evolutionary past does not lend itself to simple, linear models. The Late Pliocene witnessed a tremendous radiation of hominin populations that occupied diverse habitats, as well as the sympatric occupation by multiple species of different niches within the same habitat. Some hominin fossils from this period exhibit unexpected combinations of morphological traits, confounding attempts to place them in existing taxa.

The bottom line is that present data fail to support the notion that the bundle of hominin traits attributed to cooperative hunting evolved together as a package, or that their development was limited to a single habitat or to a single lineage of proto-human apes (Anton and Snodgrass 2012). For example, Pliocene fossil evidence now suggests that bipedalism evolved 2 million years or more before the appearance of the first humans. Similarly, tool making, once thought to be an exclusively human trait, is not only well-documented among contemporary nonhuman primates, but is indicated for some proto-human lineages as well.

Even our ideas about the relationship between brain size and intelligence may need to be re-evaluated. Ancient hominins recently uncovered at a South African cave site are a case in point (Berger et al. 2015; Berger and Hawkes 2017). The remains of several individuals, designated as *Homo naledi*, display both primitive and derived features, combining comparatively small brains with a bipedal skeletal morphology and dentition consistent with an omnivorous diet. The location and conditions under which these remains were found is suggestive of deliberate disposal of the dead and hence the possibility of symbolic behavior in the absence of significant encephalization. The recent dating of these fossils at 236 ka–335 ka (thousand years ago) poses further challenges to linear evolutionary models, since this would place *Homo naledi* as a contemporary of much bigger-brained hominins in Africa and Europe, such as *H. rhodesiensis* and *H. heidelbergensis*.

Hunting hypothesis assumptions about the role of meat and of male provisioning in the evolution of human traits have been challenged in recent years on several grounds, including interpretations of the archaeological record.[3] One of the more significant conundrums for the hunting hypothesis lies in the nature of Lower Paleolithic material culture itself. Artifacts are the archaeologist's window on prehistoric lifeways, a key element of which is subsistence activity. Food-getting and food-processing implements can say much about ancient diets. Artifacts that have survived the elements over the millennia are largely limited to impervious materials such as stone. While lithic traditions may present an incomplete picture, arguably they also reflect the focal activities of their makers. If, indeed, big-game hunting was the overarching "master behavior pattern" of proto- and early-*Homo* populations, then one would expect to find more evidence of this activity, such as weaponry, in early stone tool assemblages. But interestingly, killing implements capable of bringing down large herbivores, such as spearpoints and other projectiles, are not prominent in the paleoarchaeological record until millions of years later.

Well-documented evidence for early stone tools comes from the Afar Triangle in Ethiopia some 2.6 ma. These pebble tool assemblages, referred to as Oldowan or Mode 1, consist of crude choppers, scrapers, and cleavers. Oldowan tools may have been utilized for a time by robust australopithecines (Paranthropus), but are primarily associated with *Homo habilis* and with *H. erectus*. A refinement of Oldowan tool making, the Acheulean or Mode 2 tradition, appears by 1.76 ma in Africa, spreading to Asia, the Middle East, and Europe between 1.5 ma and 800 ka. The signature Acheulean tool type, the biface handaxe, is a ubiquitous element at *H. erectus* sites worldwide for the next million years. As such, Oldowan-Acheulean tool types represent the dominant technology for the majority of human history. It is only in comparatively recent times, with the development of Mousterian and various Upper Paleolithic assemblages, that stone hunting projectiles appear.

While wooden, bone, and fiber implements may also have played a role in the pursuit of an omnivorous diet, a reasonable argument can be made that the Early Pleistocene toolkit was principally designed for collecting and processing rather than predation.[4]

If choppers, scrapers, cleavers and handaxes were the primary tools utilized by humans for the first 2 million years of their existence, what inferences may be drawn about Early Pleistocene subsistence? These earliest stone tools have been linked to a number of food-getting activities, such as gathering root plants with handaxes or fashioned digging sticks; pulverizing seeds, nuts, and other fibrous materials with choppers; utilizing flakes for cutting animal flesh; and crushing and scraping animal bones for meat and marrow with choppers and flakes.

Hunting hypothesis advocates have eagerly pursued archaeological evidence to confirm the butchering and consumption of meat at ancient hominin sites. Artifact clusters found in association with the cut-marked bones of large and small prey have been pointed to as the smoking guns for these activities, which, it is argued, occurred at central places or base camps (Isaac 1978; Kaplan et al. 2000). Such finds, however, have been interpreted by others as more indicative of opportunistic scavenging rather than of systematic hunting and provisioning activity. O'Connell et al. (2002), in their detailed assessment of early hominin sites with bone assemblages, conclude that the procurement of meat was likely the result of active scavenging of carnivore kills. Their analysis suggests that butchery occurred near the kill sites, and that there is no evidence that animal remains were transported to central habitation places, or that meat so acquired would have been reliable or plentiful enough for the provisioning of females and young.[5]

Critics of the hunting hypothesis also point out that male provisioning behaviors and the dietary importance of hunting are not supported by observations of nonhuman primates or by contemporary hunter-gatherer societies. Although cooperative hunting does occur among male primates such as chimpanzees, there is no evidence that the spoils are transported back to a central habitation place or that such meat is shared with dependent offspring. Similarly, ethnographic data for contemporary foraging societies also fail to support the subsistence dominance of hunting (Martin 1974; Martin and Voorhies 1975: 181–82). Notably, female gathering activities are of central importance in preagricultural societies, and also prevail in contemporary foraging societies such as the Hadza of Tanzania where game has remained plentiful but where hunting is undertaken only occasionally by males, and on an individual rather than cooperative basis. More recent studies of Hadza hunting patterns confirm these findings, leading the authors to suggest that hunting may be undertaken by males for purposes other than the provisioning of family members, such as prestige and status rivalry (Hawkes 1990; Hawkes, O'Connell, and Blurton Jones 1991; Hawkes and Bliege Bird 2002).

The archaeological debate on how to interpret the faunal assemblages at early hominin sites continues. The focus of this debate is not whether meat eating occurred, but the frequency with which game was procured and the manner in which it was distributed and consumed. If, as recent analyses by O'Connell et al. (2002) suggest, the quantity of available meat was insufficient to serve as a dietary staple or as the primary nutritional basis for support of offspring, then hunting activity cannot be credited with the evolution of advanced traits that distinguish *H. erectus* from australopithecines, namely, larger body build, bigger brains, dentition changes, gut reduction, life history changes, and increases in ecological range. O'Connell et al. also question the status of meat as a literal "game changer" by citing stable carbon isotope findings (Lee-Thorp, Thackery, and van der Merwe 2000), which fail to indicate differences in the degree of dietary reliance on meat between African australopithecines and *Homo ergaster* (*H. erectus*).

Development of an alternative scenario for the appearance and longevity of *H. erectus* requires a second look at the ecological settings in which the genus evolved and prospered. The single-habitat concept centered on arid savanna grasslands has, on the basis of more complete geological and climatological data, been challenged by evidence for more diversified landscapes or mosaic habitats in areas of advanced hominin occupation. The question of *H. erectus*'s economic and social origins merits reconsideration of the nature and distribution of available resources and of characteristic primate patterns for their exploitation.

Climate Change, Water, and Ecology

African climate during the Late Pliocene-Early Pleistocene was characterized by cooler, drier, and more seasonally variable conditions. This general trend has been linked by the classic savanna hypothesis to open grassland expansion and to new selective pressures that catalyzed hominin development. More recent reconstructions, however, indicate that environments occupied by early hominins did not mirror the vast open grasslands of today's equatorial Africa, but rather consisted of highly variable, mosaic habitats. Faunal and paleoclimatic records suggest that the pace of increasing aridity at the Plio-Pleistocene boundary was not uniform, but intensified at three prominent intervals, 2.9–2.4 ma, 1.8–1.6 ma, and 1.2–0.8 ma. Each of these intervals corresponds to significant changes in faunal assemblages, including hominins (de Menocal 2004).

Based on studies of African ungulates, Vrba (1995) linked these three climate change intervals to the extinction of woodland-adapted species and their replacement by new grassland species. Her basic theory, the "turnover pulse hypothesis," is that highly specialized species are less able to adapt to changing conditions and therefore are displaced by more generalized forms. By applying this premise to early hominins, Vrba breathed new life into the savanna hypothesis with the notion that the major climatic shifts from woodland to grassland habitats marked successive stages of human evolutionary development.

Thanks to the development of refined analytical tools for paleoclimate reconstruction, a more complex picture of African Pliocene ecology has recently emerged. Areas in which early hominins lived are now known to have experienced repeated sequences of climatic instability, including wet and dry extremes, large-scale habitat oscillation, and dramatic landscape remodeling. Early hominin real estate varied in cyclical rhythms or pulses from arid to lush, from gallery forest to open woodland, and from lakes and rivers to areas of dry bushland. These findings contrast sharply with assumptions embraced by single-habitat theories. Finlayson (2014: 22–23) surveyed the features of Pliocene fossil sites and concluded that the essential habitat elements of the earliest hominins, such as *Ardipithecus ramidus*, were seasonal shallow water, trees, and open (treeless) spaces. Reed (1997) also examined habitat types associated with a variety of hominin fossils and noted that wooded, well-watered settings, such as lake margins, marshes and marshy swamps, riparian woodland, deltaic flood plains, and edaphic grasslands, are characteristic of Australopithecus and Paranthropus sites.

Potts (1996, 1998, 2013) has linked the first appearance of hominins and the evolution of genus *Homo* traits to periods of greatest habitat instability. The basic premise, known as the "variability selection hypothesis," is that bipedalism, tool making, encephalization, and expanded cognitive function arose as adaptive responses to the challenges of a diverse and rapidly changing environment. This concept was carried further by the "pulsed climate variability selection hypothesis" of Maslin and Trauth (2009), which emphasizes the influence of short periods or pulses of extreme climate change on human evolutionary development.

Research on human evolutionary ecology has now shifted to the gathering and analysis of detailed paleoclimatic records for known fossil sites. Current focus is on linking environmental changes with global and regional insolation changes at time scales of 10–100kyr. Such studies seek a better understanding of the heterogeneous spatiotemporal character of fossil site locations, and what they reveal about selective factors that may have impacted early hominin populations. Says Kingston:

> Although the potential significance of these climatic changes has been long recognized, only recently has the resolution of investigations allowed specific detection of the effects of this cycling at early hominid fossil localities. Emerging data indicate that insolation changes occurring at the precessional frequency (23ka) had significant effects on early hominid habitats, with magnitudes of change similar to that proposed for intervals spanning millions of years. Climatic variability associated with these changes resulted in persistent ecological change in equatorial Africa in which communities were continually fragmented and reassembled in novel ways. (2007: 58)

Dominguez-Rodrigo (2014) argues that, when viewed from the broader ecological perspective of "biomes," recent evidence for the presence of diversified or mosaic habitats does not negate the basic assumptions about hominin evolution advanced by the original savanna hypothesis. The term "savanna," he maintains, is indicative of a biome type that is associated with a grasslands substrate but that may also contain a variety of habitats including woodlands and open forests. Thus, while acknowledging that the Pliocene was more humid than previously envisioned, and that the transition from dense forest to open grasslands was probably more gradual, Dominguez-Rodrigo holds that the basic evolutionary scenario presented by the savanna hypothesis is still viable.

What the savanna hypothesis fails to anticipate, however, is the potential evolutionary significance of well-watered habitats. Of particular interest for the reconstruction of Late Pliocene ecology are recent core samplings of ancient lake beds in the East African Rift Valley. Gibbons (2013) summarized the findings of lake sediment studies for ten rift basins, noting that at least three humid periods occurred over the past 8 million years, each associated with the filling of deep lakes and the appearance of new hominins. Similarly, a study of the Olduvai lake basin by Ashley (2007) indicates that the lake expanded and contracted five times between 1.85 and 1.74 ma in a 21-ky cycle in which rainfall is projected to have increased by one-third between wet and dry seasons. Ashley sees a potential link between these cycles, hominin evolution, and early *Homo* migrations out of Africa around 1.8 ma. A comparable 20-ky cycle of lake filling and emptying is noted by Gibbons for the Baringo Basin, along with *Homo* fossil remains dated 2.5 ma.

Joordens et al. (2011), in their investigations of the fossil-rich Turkana Basin, highlight the potential implications of paleo-lakes for ancient hominin ecology:

> Our results demonstrate that between ~2 and 1.85 Ma the Turkana Basin remained well-watered and inhabited by hominins even during periods of precession maxima when summer monsoon intensity was lowest. This is in contrast to other basins in the East African Rift System (EARS) that were impacted heavily by precession-forced droughts. We hypothesize that during lake phases, the Turkana Basin was an aridity refugium for permanent-water dependent fauna—including hominins—over the precessional climate cycles. (2011: 1)

Analyses by Maslin et al. (2015) have statistically correlated hominin speciation events, such as the emergence of *H. erectus*, with periods of maximal East African Rift System paleo-lake coverage, and also suggest a link between fluctuating lake levels and hominin dispersals. The vast interlacustrine system of the East African Rift Valley played a critical role in both the evolution and periodic migrations of hominins in the Late Pliocene, providing both resource-rich habitats and, in some instances, refugia from the more extreme swings in climate cycles.

Don't Go Near the Water

While recent reconstructions of African paleoecology have documented the heterogeneous nature of ancient hominin landscapes, site-specific studies have also called attention to one common feature they share across a broad spatiotemporal spectrum—proximity to water. Until recently, the cyclical availability of well-watered habitats in the Late Pliocene has received comparatively little attention in academic circles. Debate has focused, instead, on either proving or disproving the alleged relationship between grasslands expansion, bipedal origins, and the primacy of hunting. Even in the face of new information, the basic underlying assumptions of the savanna hypothesis have been amazingly resilient.

Thus, when prominent African archaeologist Desmond Clark (1980) took a step back from the single-habitat theory by acknowledging the presence of mosaic habitats at the Plio-Pleistocene boundary, neither he nor other savanna hypothesis supporters were motivated to reconsider other contingent assumptions about early human evolutionary ecology. So entrenched was the portrait of early man hunting on the arid plains that the connection of ancient fossils to water sources was dismissed largely on the basis of hydration requirements or the lure they presented to unsuspecting prey. Consequently, consideration of well-watered locations as permanent or semipermanent habitation areas during critical periods in hominin development has been largely ignored. Similarly, exploring the potential ramifications of such aquatic adaptations for subsistence and for the evolution of human morphological and behavioral traits has, historically, been a place where untenured professors and graduate students dared not go. Such ideas and lines of scientific inquiry have, in fact, been met with both cursory dismissal and even ridicule.[6]

To understand why this is so, one has to appreciate the academic political climate of the 1970s. This was a time when the hunting hypothesis had crystallized to the level of accepted dogma in paleoanthropology, along with its gender-biased assumptions about human social origins. Sociobiology also emerged as a major body of theory during this same time frame, the initial iterations of which viewed human society, past and present, through an arguably androcentric lens. Dissemination of these ideas in the popular press coincided with the rise of the feminist movement, often stirring the angst of its leaders, who hurled accusations of sexism and biological reductionism on the architects of social origins theories. It was a time when academicians sought to draw the line between "scientific" and "populist" discourse, a line that scholars—and particularly women scholars—crossed at their own peril.

Into this rather volatile intellectual climate stepped academic outsider Elaine Morgan, whose popular 1972 best seller *The Descent of Woman* directly challenged establishment theories on human origins. Drawing on earlier observations made by Oxford marine biologist Sir Alister Hardy (1960), Morgan proposed that our human ancestors occupied an aquatic rather than savanna grasslands habitat during a critical period of their evolution. This semi-aquatic existence, it was argued, gave rise to a number of morphological and functional traits, such as bipedalism, hairlessness, subcutaneous fat, and voluntary breath control, that collectively make our species unique among apes and most other animals. Morgan refined and expanded on this theory, known as the "aquatic ape hypothesis" (AAH), in three subsequent books, *The Aquatic Ape* (1982), *The Scars of Evolution* (1990), and *The Aquatic Ape Hypothesis* (1997).

Despite wide dissemination of Morgan's books, the mainstream scientific community remained unreceptive and it would be several years before elements of the AAH would receive serious consideration (Ellis 2011; Kuliukas 2011; Williams 2011). A primary reason cited by evolutionary theorists for dismissal of her work was the perception of *The Descent of Woman* as a feminist treatise. Morgan's at times polemical tone and unabashed attacks on prevailing androcentric theories colored the reception of her ideas among scholars. This book and her subsequent works were summarily dismissed as catering to lay rather than scientific audiences. Morgan's background as a television scriptwriter and lack of formal academic credentials also contributed to portrayal of the AAH as popular fiction. Kuliukas (2011) argues that the sole peer-reviewed critique of the AAH, written by Langdon (1997), misrepresented the primary thrust of Morgan's thesis and conducted neither a comprehensive nor unbiased review of its basic assumptions. Still, this lone critique was sufficient to silence academic discourse on the AAH, perhaps discouraging the very scientific research necessary to assess its validity and its general usefulness as a framework for unraveling the human origins puzzle.

A significant turning point occurred in 1995 when, during his Daryll Forde Memorial Lecture at University College, London, prominent African paleontologist Phillip Tobias (1998) publicly abandoned the original savanna hypothesis, reversing his former position of support. Tobias noted in retrospect that he was increasingly persuaded by evidence linking East and South African fossil sites with forested and forest margin habitats and by five critical ways in which water and the proximity to water may have influenced the course of human evolution, namely: "drinking, keeping cool, anthropogeographical dispersal, aquatic adaptations, and aquatic foods" (2011: 13). The reversal was an act that Tobias personally credited with removing objections to the AAH on its face, thereby encouraging a rigorous multidisciplinary examination of its basic tenets.[7] With what Tobias called the "liquidation" of the savanna hypothesis, the gates of scientific inquiry were effectively opened to alternative models of human evolutionary ecology.

Now that Late Pliocene hominins are known to have occupied well-watered habitats at critical periods in their evolutionary development, new questions arise. What resources would have been available to them in these settings? How did the waxing and waning of paleo-lakes and river systems impact the subsistence, settlement patterns, migrations, and extinctions of early terrestrial primate taxa? What roles did aquatic flora and fauna play in the evolution of omnivorous diets, and what are the implications of their exploitation for early *Homo* morphology and encephalization?

Semi-Aquatic Habitats, Resources, and Dietary Lipids

The fossil remains of aquatic animals have been long recognized as a component of early hominin sites, but their dietary significance has been largely overlooked until recent years. Stewart's (1994) seminal study of fish as a staple food in early hominin diets analyzed several Olduvai Gorge sites at which sizeable *Clarius* (catfish) and *Ciclidae* (tilapia) bone assemblages have been found. Stewart observed that these species may have played a central role in the subsistence of ancient as well as modern peoples due to their high nutritional value, abundance, seasonal spawning and nesting patterns, and the ease with which they may be procured:

> The optimal periods and locales for procuring fish with minimal or no technology are therefore river floodplains in the early long wet season (usually November-February in eastern Africa), when waters recede and fish are trapped in isolated pools. Mainly cichlids and catfish, but also other inshore taxa and young fish are trapped at this time. The late dry/early wet season would presumably be of most importance to hunter-gatherer groups, as this is when fish are generally in peak condition; this also coincides with the period of poorest nutritional value of plants and terrestrial mammals. (1994: 233)

Stewart views early hominin sites with an abundance of fish remains as comparable to better-known Late Pleistocene riverine and delta occupations that relied on the procurement of seasonally spawning taxa on a multiyear basis. Unlike the scavenging of carnivore kills, regular procurement of fish would provide an abundant and reliable source of high-quality protein with minimal risk.

At the time of Stewart's study, the association of Oldowan tools or hominin remains with fish and other water-dependent genera was known, but hard evidence for their consumption (i.e., cut-marked bones) was lacking. Subsequent discoveries, such as the 1.95 ma site in the Lake Turkana basin (designated as FwJj20), removed all doubt (Braun et al. 2010; Steele 2010). This site, located in a riverine forest habitat at the time of occupation, contains Oldowan tools and the butchered remains of both terrestrial and aquatic animals. It provides definitive evidence for the exploitation of catfish, turtles, and crocodiles, along with the bones of larger, likely scavenged animals, including hippopotamus, rhinoceros, and bovids:

> The evidence from FwJj20 indicates that hominins were very effective at securing access to a wider variety of high-quality animal tissue than has been previously documented. Some of these resources would have provided necessary dietary resources without the added predation risks associated with interactions with large mammalian carnivores that are sometimes involved with the acquisition of elements of large animal carcasses.... In addition, although animal tissues provide nutrient-rich fuel for a growing brain, aquatic resources (e.g. fish, crocodiles, turtles)

are especially rich sources of the long-chain polyunsaturated fatty acids and docosahexaenoic acid that are so critical to human brain growth. . . Therefore, the incorporation of diverse animals, especially those in the lacustrine food chain, provided critical nutritional components to the diets of hominins before the appearance of *H. ergaster/erectus* that could have fueled the evolution of larger brains in late Pliocene hominins. (Braun et al. 2010: 10005)

This Lake Turkana site indicates that some hominins had already committed to a fully omnivorous diet by 1.95 ka, including reliance on local aquatic resources. It also suggests that these hominins may have begun to partition the habitats they shared with robust australopithecines, such as *Paranthropus boisei,* carving out for themselves a niche that provided dietary flexibility in the face of changing climate conditions. In so doing, they may have gained an adaptive advantage over their sympatric neighbors. It also set in motion a series of rapid morphological and behavioral changes that would define the genus.

Several features that distinguish *H. habilis* and *H. erectus* from australopithecines have been linked to dietary shifts and nutritional sufficiency (McHenry and Coffing 2000). These include an increase in body size, a decrease in gut size, changing dentition, and increases in both energy storage (adipose tissue) and energy expenditures related to foraging activities (Leonard and Robertson 1998; Milton 1999; Anton 2012; Ungar 2012). As noted above, the incremental expansion of hominin brains has also been tied to nutritional changes involving the dietary consumption of long-chain polyunsaturated fatty acids (LC-PUFA) and docosahexaenoic acid (DHA), both known to be critical to the metabolically expensive processes of brain growth and function.

The DHA content of marine fish and shellfish is five to one hundred times that of lean and fatty meat, while marine bird egg yolks and mammalian brain tissue also constitute rich terrestrial LC-PUFA sources. The high concentration of LC-PUFA and DHA in the littoral and marine food chains have led several investigators to state that hominin encephalization was inexorably linked to and dependent on the consumption of aquatic animals.[8] Broadhurst et al. note:

Exploitation of river, estuarine, stranded and spawning fish, shellfish and sea bird nestlings and eggs by *Homo* could have provided essential dietary LC-PUFA for men, women, and children without requiring organized hunting/fishing, or sophisticated social behavior. It is however, predictable from the present evidence that exploitation of this food resource would have provided this advantage in multi-generational brain development which would have made possible the advent of *H. sapiens.* Restrictions to land based foods as postulated by the savannah and other hypotheses would have led to the degeneration of the brain and vascular system as happened without exception in all other land based apes and mammals as they evolved larger bodies. (2002: 653)

As noted earlier, Maslin et al. (2015) have correlated hominin speciation and encephalization with the appearance of East African Rift deepwater lakes. Such bodies of water could have provided the essential habitat for a variety of aquatic fauna that fed brain growth.

Despite cumulative evidence pointing to the importance of aquatic resources in the emergence of early *Homo*, the hunting hypothesis and its corollary assumptions have not been laid to rest. In *Catching Fire: How Cooking Made Us Human*, Wrangham (2009) resurrects meat eating and male provisioning as central themes in human evolution, but with a nuanced twist. Namely, he proposes that while meat provided the initial nutritional boost that elevated habilines (*H. habilis*) over the australopithecines, it was the cooking of meat that accounts for the emergence of *H. erectus*. The use of fire for food preparation is seen as a critical step in making meat softer and more digestible, thereby increasing its caloric and energy values. Wrangham sees this nutritional increase as pivotal since, he claims, valuable energy sources such as lipids were unavailable to ancient hunter-gatherers (2009: 21).

An obvious shortcoming of Wrangham's argument is that lipid-rich aquatic resources were indeed available. The primal habitats of early humans contained abundant sources of dietary lipids, and a variety of such water-dependent genera are now known to have been regularly procured. Many of these sources, such as fish, were soft and easily digested foods that could have been consumed without cooking. Moreover, consumption of these aquatic foods in the Late Pliocene precedes the first undisputed evidence for fire use by about a million years. Although recent sites in South Africa suggest a much earlier origin, the systematic cooking of foods by *H. erectus* remains in question.[9]

Wrangham goes on to propose that the advent of cooking also accounts for the evolution of early human domestic units and the division of labor by sex. Male-female pair bonding, he argues, was not forged on the basis of sexual competition. Rather, for Wrangham, it all boils down to mastication. The chore of chewing raw meat, he notes, was extremely time-consuming and took away valuable hours from male hunting activity. Fire provided a solution by reducing chewing time, allowing hunters to spend a full day at the hunt and return to camp at nightfall with time to spare for an evening meal.

Wrangham theorizes further that females were forced into domestic roles by patriarchal males who relegated their mates to "keeping the hearth" in exchange for protection against other males who would seek to steal their food. In a manner reminiscent of earlier man-the-hunter models, the cooking hypothesis rests on assumptions of male provisioning and dietary reliance on meat, neither of which are supported by recent evidence. This scenario also deviates sharply from Wrangham's (1979, 1980) earlier works on primate ecology discussed in chapter 2, which concluded that female subsistence and mating strategies exert a primary influence on the structure of primate socioeconomic groups. As noted by O'Connell et al.:

> If big game hunting, central place foraging, and paternal provisioning are eliminated from the *H. erectus* mix, then support for contingent inferences about modern human-like social organization and mating arrangements disappears. Reconstructions of its behavior are reduced to intriguing but somewhat disconnected inferences based on skeletal anatomy and archaeology; questions about its evolutionary origins and persistence are largely begged. (1999: 462)

An Alternative Plio-Pleistocene Scenario

One of the inherent dangers in tracing the emergence of the genus *Homo* is the tendency to create ancient hominins in our own image. As we have seen, models of early human society have historically reflected Western European cultural values on the nature, aptitudes, and juxtaposition of the sexes. These biases have been reflected in the interpretation of paleontological and archaeological data, as well as in the projected social implications of material remains. In constructing an alternative scenario on human origins, then, it is perhaps prudent to begin with the cautionary conclusion reached by O'Connell et al., namely: ". . . early humans were behaviorally unlike moderns" (2002: 862).

What can be reasonably proposed about the socioecology of early human primates living in the East African Rift Valley in the Late Pliocene? A number of initial conclusions may be drawn based on discussions in this and preceding chapters. We begin this scenario with the following assumptions: (1) male and female reproductive strategies in early human lineages mirrored those of contemporary primates, namely, male inclusive fitness was based on the number of offspring produced, and hence access to fertile females, whereas female inclusive fitness was based on access to sufficient food for the successful nurturance and provisioning of offspring from zygote to maturity; (2) the basic early hominin socioeconomic unit was the matricentric family, consisting of a mother and her dependent offspring, (3) female feeding strategies and the sociospatial distribution of matricentric units were optimized in relation to the distribution of resources in a given niche; (4) the socio-spatial distribution of males was related to, and in large part dictated by, the distribution of females; (5) the mosaic habitats comprising early hominin occupation areas offered high-quality food resources distributed in discrete, defensible patches, often spread over broad ranges and seasonally variable; and (6) as among contemporary primates living in such habitats, the predominant social organization of proto-humans likely consisted of nonterritorial, permanent, often multigenerational female-bonded kin groups, with multiple males selected for group membership on a competitive basis.

This scenario assumes that our proto-human ancestors were members of intelligent ape lineages that gradually dispersed into well-watered forest fringe habitats in response to climate change events. Consistent with Tanner's (1981) gathering hypothesis, the vanguard of this movement is proposed to have consisted of groups of related females and their offspring who successfully expanded their dietary repertoire from fruits and berries to a variety of roots and tubers, as well as protein sources such as nuts, insects, eggs, lizards, turtles, and small mammals. Early tool use likely evolved as an innovation for intensive food gathering, expanding from collection to processing—an important component in food preparation for altricial young. Food sharing had its origins in maternal investment in offspring and was later extended to other proximate lineal and collateral kin.

Notably, in this scenario of early human social life, both tool use and food sharing innovations debuted in the absence of organized hunting or male provisioning. While males also gathered, they likely did so largely for their own subsistence or to gain acceptance into non-natal female-bonded groups. Tanner notes how sexual selection of "friendly" males may have dovetailed with natural selection over time to reduce dimorphic traits within the taxa. Males may have also engaged in aggressive scavenging, but primarily as an activity of intrasexual competition and display rather than of meat provisioning of purported offspring (Hawkes and Bliege Bird 2002; O'Connell et al. 2002).

After millennia of successful adaptations and radiations, gracile australopithecine lineages, such as *A. afarensis*, disappear from the fossil record by 2.8–2.5 ma, while robust lineages, such as *Paranthropus boisei*, persisted alongside more advanced hominins for another million-plus years (Reed 1997: 316–17). Stanley (1992) hypothesizes that since gracile lineages lived in rather static woodland and riverine forest habitats, they may have fallen victim to increasing aridity, whereas robust varieties survived considerably longer by adapting to life in edaphic grasslands.

An alternative hypothesis is that some lineages of gracile australopithecines or other early hominins, rather than succumbing to extinction, adopted innovative subsistence and reproductive strategies that catalyzed their evolution into habiline and early *H. erectus* forms. The differences between australopithecines and early human species, however, are notable. *H. erectus* was, on average, about 30 percent larger in body size. Comparisons between *A. afarensis* and early African *H. erectus* reflect a similar percentage disparity in cranial capacity (Anton 2012; Anton and Snodgrass 2012; Pontzer 2012). Although nominally omnivorous, australopithecines also retain a broad pelvic girdle, noted by some as indicative of a sizeable gut and primary dependence on vegetative foods. In sum, the morphology of australopithecine taxa reflects a brain size and life history similar to modern chimpanzees, with early maturation and a late age at weaning. This suggests only limited provisioning of offspring, and the absence of extended postmenopausal lifespans associated with intergenerational child nurturance (O'Connell et al. 2002: 834).

What chain of events could have precipitated the relatively rapid evolution of bigger brains and bodies and the broad spectrum of behavioral and life history changes that marked the emergence of *H. erectus*? Prevailing theory is that advances culminating in the genus *Homo* are rooted in the ability of some hominin lineages to solve the climate change riddle. This solution, it seems, had two essential and interrelated components, both of which are correlated with increasing encephalization. The first strategy was ecological. Successful proto-humans developed a mastery of their natural environment and the spatiotemporal patterns of its landscapes, life forms, and resources. This road-mapping ability, perhaps honed by their forebears in forest settings, was utilized not only to expand the scope of available high-quality foods, but to enhance the efficiencies by which they were procured. The second essential component was a fundamental shift in reproductive strategies. Cooperative breeding reset the energy budgets and population growth profiles of early human groups, leading, in turn, to dramatic changes in *Homo* life history traits. Females played a pivotal if not primary role in reshaping both the subsistence and reproductive strategies critical to hominin advancement.

Female Subsistence Strategies

As the species' principal provisioners, females were in a perpetual food quest for themselves and their offspring. They were therefore on the front lines in the battle to satisfy the next generation's nutritional needs in the face of changing landscapes and conditions. As among contemporary female primates, the intrasexual hierarchies of female-bonded groups also likely served to regulate variable access of their members to high-quality foods, as well as pacing group movements in relation to resources.

O'Connell et al. (1999) proposed that climate-driven changes in female foraging patterns and food sharing laid the foundation for hominin advancement. The focus of their study was on the exploitation of tubers, or underground storage organs (USOs), that were both abundant and a rich source of carbohydrates. Other investigators have highlighted the nutritional ramifications of body and brain size increases in early *Homo,* suggesting that an even broader range of high-quality foods may have been required. Put simply, greater size translates into elevated total daily energy expenditures (TDEEs) and the need for enhanced energy inputs.

Assuming a physical activity level (PAL) comparable to that of contemporary apes, Foley and Lee (1991) estimated TDEEs for *H. erectus* to be 40–50 percent greater than that of australopithecines, and up to 80–85 percent greater if subsistence activity levels of early *Homo* were comparable to that of modern hunter-gatherers. Anton and Snodgrass (2012) were more conservative, estimating TDEE disparities between *H. erectus* and *A. afarensis* at only 15 percent if ape PALs were assumed, but predicted much greater differences if human-like PALs prevailed. Aiello and Key (2002) focused on comparative energy estimates for gestating and lactating females, concluding that *H. erectus* life history changes accounted for TDEEs 47 percent greater than those of robust australopithecines. While such estimates vary, investigators are in agreement that *H. erectus* increases in body and brain size had a significant impact on daily metabolic costs, requiring enhanced energy inputs for maintenance, growth, and reproduction.[10]

Increased body size, changes in dentition (reduction in the size of cheek teeth), and gut size reduction reflect an increasing reliance on animal foods. *H. erectus* individuals of both sexes were not only eating greater quantities, but likely relied on a diet that was rich in essential fatty and amino acids. As pointed out by Milton (1999), increases in body size among mammalian herbivores is generally associated with a decrease in the quality and energy output of food consumed. Gorillas, for example, must eat large amounts of vegetative matter daily to support the minimum energy needs of their substantial biomass. By shifting to an omnivorous diet rich in amino acids and other essential micronutrients, *H. erectus* was able to consume smaller volumes of food for greater energy returns, achieving body size increases while avoiding the typical mammalian limitations. But were gatherable sources of such high-quality foods, such as animal protein, readily available to female hominin providers?

There has been a recent resurgence of interest in the relative importance of male subsistence contributions to matricentric units in the form of hunted or scavenged meat. A central question is whether alloparental care (grandmothering) and female provisioning alone could support the increased energy requirements of evolving hominins. Kaplan and colleagues (2000), for example, essentially resurrected the hunting hypothesis by crediting paternal recognition, male parental investment, and meat provisioning of dependent females and offspring with life history changes essential to brain expansion. Others have pointed to the importance of meat sharing as a risk reduction factor in the nutrition of contemporary hunter-gatherer groups (Layton 2008: 123–24).

Opie and Power (2008) recently revisited this issue by examining the life history and production versus consumption patterns of contemporary chimpanzees and human foragers. These data were then utilized to create a model of *H. erectus* energetics, with the assumption that early hominins represent an intermediate point between the two species. Given the projected availability of adult helpers and the estimated caloric value of food sources, they concluded that the hominid to hominin transition would probably have required females to enlist the support of both their female kin and cooperative males. Intersexual alliances for the sharing of scavenged meat, however, may emerge in the absence of pair bonding and male retention. Indeed, Opie and Power conclude that only a social organization based on female philopatry could have provided the essential structure for both allocare and supplementary caloric contributions from friendly males.

Notably, calculations of food caloric values in the Opie and Power study are based in large part on average production versus consumption ratios recorded for contemporary hunter-gatherer societies, the bulk of which live in marginal habitats. An important question, therefore, is the extent to which these habitats are typical of those in which *H. erectus* evolved. Namely, are they representative of the mosaic landscapes that may have provided substantially different opportunities affecting the type and density of resources, food-getting energy expenditures, and caloric returns?

Historically, procurement of animal protein and provisioning with high-quality foods has been cast outside the female domain. But it is entirely reasonable to assume that a large variety and quantity of animal foods were exploited by Plio-Pleistocene females, both individually and collectively, as a natural extension of intensive gathering activity. Procurement of protein sources such as seeds, nuts, and small game are frequently mentioned in current models (Jolly 1970). Seldom acknowledged, however, are the abundant sources of animal protein, such as fish and mollusks, that were readily available in aquatic food webs. Exploitation of similar aquatic resources by nonhuman primates is well-documented. South African baboons, for example, have been observed beachcombing for shellfish and shark eggs. Similarly, female long-tailed macaques in coastal Thailand regularly gather crabs and oysters and utilize stone tools to crack open oyster shells and to scrape mollusks off of rocks (Malaivijitnond et al. 2007; Broadhurst, Crawford, and Monroe 2011).

Early humans were no strangers to water and were likely adept at wading, swimming, and diving as routine food-getting activities. Ethnographic examples of harvest diving for shellfish and other underwater flora and fauna are recorded as an ancient subsistence activity among several Southeast Asian peoples. Among the Ama of Japan, for example, active divers are predominantly women over the age of sixty, with the oldest reported at ninety-three (Schagatay 2011). This type of food-getting activity by "grandmothers" would be potentially valuable to any cooperative breeding population living in proximity to bodies of water. Rivers, lake shorelines, deltas, and estuaries are also home to abundant tubers, fruits, berries and other nutritional foods that could be gathered within easy range of encampments.

As discussed above, Stewart (1994) notes that large volumes of fish could have been harvested in East African paleo-lakes and river shores on a predictable basis. Moreover, they could be gathered largely by hand without any special technology, and by most community members irrespective of sex and age. If, as Stewart claims, one 40 cm tilapia fish could feed a family of four, it is difficult to support any model of *H. erectus* subsistence that excludes aquatic protein sources as important dietary staples. Broadhurst et al. observe:

> Rather than running over open plains or relying on others for their provisions, nursing and pregnant women, the primary providers of nutritional needs for the developing generation, could have collected all the nutritious food they needed by inhabiting the water's edge. Likewise for children and the elderly, injured and sick. Prodigious amounts of LC-PUFA can be obtained with little effort or organization by collecting shell fish, turtles, lizards, snakes, frogs, and eggs of flightless birds. (2011: 20)

Collection of aquatic resources such as mollusks and fish would have provided a reliable alternative to mammalian meat as well as a buffer against food shortages experienced during the most arid calendar phases. Stewart observes that both tilapia and catfish attain their

Figure 3.1. *Homo erectus* lakeshore encampment in the Early Pleistocene. By Drew Fagan.

best condition (and hence their highest fat content) prior to spawning at the end of the dry season. Harvesting at this time would coincide with the period when both vegetative and mammalian meat resources were the least plentiful and at their poorest nutritional value.

The chemistry of aquatic resources is credited with two additional elements essential to the emergence of *H. erectus*—their suitability as brain food and as baby food. As noted earlier in this chapter, fish and shellfish contain a lipid profile (LC-PUFA, DHA) regarded as essential to both human brain growth and hominin brain evolution. Additionally, the lacustrine food chain would have provided lipid-rich protein sources in the form of soft foods suitable for consumption by weanlings—a critical piece in the evolving life history profile of early humans.

Female Reproductive Strategies

The second component central to the emergence of *H. erectus*, namely, a shift in reproductive strategies, likely occurred in tandem with climate-induced changes in female foraging patterns. Indeed, female inclusive fitness, by definition, relies on the interdependence and parallel optimization of subsistence and reproductive behaviors. Females who shared food and childcare responsibilities with close kin enjoyed better nutrition, greater infant survival rates, shorter birth intervals, enhanced fecundity, and longer lifespans. Female-bonded kin groups that structured access to both resources and mating opportunities established stable multimale-multifemale communities that advanced the fitness of both sexes.

The net effect of cooperative breeding on the energy expenditures and fecundity of females was far-reaching. At the individual level, foraging success for *H. erectus* females was limited by climate factors that affected the distribution of high-quality foods, as well as the increased energy expenditures required for their collection. Access to preferred foods was also likely affected by the position of one's natal matricentric unit in the hierarchy of female-bonded kin groups composing the local community. These limitations could have been mitigated by the sharing of both food and childcare responsibilities among close female kin such as sisters and grandmothers.[11] Females may have also successfully reduced their energy expenditures by utilizing their expanded sexual receptivity to solicit cooperation from males within the polygynandrous mating population. According to a model developed by Key (Key 1998; Key and Ross 1999), males are predicted to engage in nonreciprocal altruism when there is a 50 percent or greater chance that such cooperation will result in sexual access.

The interrelationship between subsistence and reproductive strategies is also reflected at the group or community level. Ongoing access to high-quality foods on a regional basis was likely influenced by intergroup (and perhaps inter-taxa) relations. As noted for contemporary primates, dominance relationships among neighboring nonterritorial female-bonded groups is linked to group size, with larger groups and groups with multiple adult males prevailing (Wrangham 1980). The selection and adoption of additional males into established female-bonded groups would have provided defensive advantages for maintaining access to discrete patches of preferred foods. This strategy may have been utilized by *H. erectus* for a similar purpose, perhaps foreshadowing the partitioning of niches. Increased male membership in *H. erectus* communities provided new opportunities for intersexual economic cooperation, thereby reducing the energy load for females. It also greatly expanded the mating opportunities for males in the breeding population, thereby enhancing their inclusive fitness and the genetic diversity of local groups.

Encephalization and Energetics

Of all the side-by-side comparisons of *H. erectus* with australopithecines and other early hominins, the most compelling lies in increasing brain size. As measured in geological time, the pace of Late Pliocene encephalization was very rapid indeed. An estimated 20 cubic centimeters of brain tissue was added every 100,000 years, or approximately every 5,000–5,500 generations (Boaz 1997: 140–42). Recent human origins theories have suggested that selective pressures for the evolution of big brains and the honing of cognitive abilities arose in response to survival challenges presented by climate instability. But what were the underlying conditions that set this process in motion for *H. erectus*? And, conversely, what factors account for brain size limitations in sympatric taxa such as *Paranthropus boisei* or in contemporary nonhuman apes?

Encephalization is an autocatalytic product of the unique alignment of increased foraging efficiencies, high-quality foods, and changes in reproductive strategies. One body of theory has focused on the energetic consequences of these changes for *H. erectus* females. Big brains take a lot of energy to grow and maintain. Since the costs of encephalization are borne by females, both in utero and during lactation, the evolution of big brains must be compatible with changes in maternal energetics (Foley and Lee 1991). The high energetic burden of reproduction was offset by females in two ways: (1) by increasing the quantity and quality of available energy inputs, and (2) by reducing energy expenditures through the enlistment of reciprocal cooperation and assistance of close kin for foraging and allocare.

To meet the expanded energy needs of themselves and their infants, *H. erectus* females broadened their range of foods to include tubers rich in antioxidants as well as abundant sources of dietary lipids and protein, such as fish, shellfish, turtles, snakes, and bird eggs. Aiello and Wheeler (1995) have argued that brain tissue is metabolically "expensive," and that these new dietary strategies offset the energy expenditures of larger brains by simultaneously reducing the size of other expensive bodily tissue, such as the gut. Nutritional enrichment was also accomplished by cooperative foraging and food sharing. It is reasonable to assume that *H. erectus* females enhanced their subsistence efficiencies by engaging in collective food-gathering activity, particularly during periods of abundance such as fruit ripening or fish spawning. Other foraging endeavors, such as the driving and ambush of small animals, could have involved the collective efforts of both adults and young children who shared food together.

But perhaps the most effective strategy employed by *H. erectus* females to reduce their reproductive energy burden was to garner the assistance of their mothers and other proximate female relatives for the mutual feeding and care of infants and weanlings (Foley and Lee 1991; Hawkes et al. 1998; Aiello and Key 2002). Through alloparenting, grandmothers not only enhanced their daughters' reproductive success, but also their own inclusive fitness by promoting and preserving their shared genes in the descendant generation.

Cooperative breeding has been hailed as the cornerstone of critical energy and life history changes necessary for hominin encephalization. Isler and van Schaik (2012) argue that the relationship between energy investments, mortality rates, and reproductive rates places an inherent limit on brain expansion. Their so-called "gray ceiling" theory proposes that increases in brain size are accompanied by increased lifespans that cannot compensate for the energy costs of slow development and lower reproductive rates. As a population reaches its minimal mortality level, brain size hits a ceiling beyond which extinction occurs. The authors develop a mathematical model that concludes that cooperative breeding and the accompanying redistribution of available energy to mothers and infants is the only way that population growth may be sustained for brains larger than 700 cubic centimeters. As would be predicted by this model, smaller-brained hominins such as australopithecines exhibit none of the characteristics that suggest cooperative breeding, such as delayed maturity, extensive childhood provisioning, and postmenopausal longevity (Hawkes et al. 1998). Similarly, the life history profiles of contemporary nonhuman apes such as chimpanzees are comparable to those of ancient hominins that remained below the gray ceiling.

As discussed earlier, Tanner (1981) linked sexual selection by female members of polygynandrous groups and the declining importance of male intrasexual competition for group access to the reduction of hominin sexual dimorphism. More recent primate studies on the relationship between sexual dimorphism, mating success, and neocortex size have supported this notion. Pawlowskil, Lowen, and Dunbar (1998) found that relative neocortex size rather than a male's rank in the dominance hierarchy was a better predictor of mating success. They concluded that males with large neocortices relied more on their brains than brawn to avail themselves of sexual opportunities with estrus females. Parallel conclusions were reached by Schillaci (2008), whose study of thirty primate species found that the intensity of male competition (as reflected in the level of sexual dimorphism) was inversely related to neocortex size (but see the discussion of mosaic brain evolution in chapter 5).

Early forms of australopithecus, such as *A. afarensis*, are often noted as exhibiting stronger dimorphic traits than later australopithecines or *Homo*. Preliminary conclusions drawn on the range of variation in body and dental morphology within and between ancient species and between males and females are necessarily tentative due to the difficulty of sexing incomplete fossil remains and to small sample sizes (Anton and Snodgrass 2012; Plavcan 2012). Nonetheless, there is an obvious macro-trend in human evolution toward the reduction and eventual disappearance of robust dimorphic traits, such as large honing canines, associated with male sexual competition and aggression.

Indeed, it is the absence of such traits in early bipedal hominins such as *Ardipithecus ramidus* (White, Suwa, and Asfaw 1994) and the persistence of more primitive dental characteristics among subsequent australopithecine taxa that convinced Lovejoy (2009) to reject the latter as an intermediate form between ancient ape-like ancestors (chimpanzee-centric models) and humans. Notably, Lovejoy sees contemporary African apes as an improbable model for ancestral hominins, since their evolutionary trajectory was distinct and has resulted in highly derived characteristics based on sexual agonism and contrastive dietary and locomotor patterns. Similarly, australopithecines, although bipedal and nominally omnivorous, retain prominent canines and ape-like life history patterns. *A. ramidus* populations are of interest in that they, and perhaps other as yet undiscovered ancient hominin lineages, exhibit none of the dimorphic traits associated with male intraspecific aggression, suggesting that the so-called emancipation of primitive primate reproductive behavior may have begun much earlier than current phylogenetic models propose.

To summarize, the alternative Plio-Pleistocene scenario presented here considers ancestral hominins first as mammals, then as primates, and finally as *human* primates. This perspective highlights the pivotal role played by female inclusive fitness in the human evolutionary process. As mammals, females carry the sizeable energy burden of reproductive nurturance, from zygote to maturity. They are the natural providers, the initial food sharers, and the agents of early socialization for the species. As primates, they pursue these responsibilities in characteristic ways, forging social and economic relationships within the breeding population that optimize their success in relation to available resources. Entry into new mosaic habitats in the Late Pliocene greatly expanded the range of high-quality foods available to female foragers. The spatial distribution of these foods in prime habitats selected for the type of primate social organization that clustered female-bonded groups. These uterine kinship units structured access to resources through intrasexual ranking systems and the orchestration of group movements. They also served to regulate sexual activity through the selective inclusion of adult males in group membership, and the maintenance of polygynandrous mating relationships. Finally, as human primates, females successfully reduced their growing reproductive energy burdens by enlisting the cooperation of close lineal and collateral kin for the feeding of themselves and their altricial young. Such cooperative breeding strategies redistributed the energy budget available to local groups and set the stage for encephalization and the expansion of executive regions of the brain.

This scenario is a notable departure from traditional theory not only in the evolutionary role it assigns to females, but in the way in which it treats male inclusive fitness. The key driver of male reproductive success is access to fertile females for the siring of offspring. Sexual access for male primates is regulated by both intrasexual competition and female selection. Among contemporary primates with polygynous mating systems, sexual access is limited to a very small percentage of males—a privilege earned by their position in the male hierarchy plus their ability to attract and retain females with their margin of safety. Where polygynous mating prevails, therefore, the relationship of adult males with female domestic units is largely peripheral. In contrast, resources in more open mosaic habitats are distributed over an area too large for a single male to defend against rival conspecifics. Under these conditions, groups are often nonterritorial and the criteria for sexual access to females change accordingly. Males may still compete with one another for favored status, but participation in multiple polygynandrous unions is newly available to selected males whose group membership provides a measure of security in intra- and intergroup relations. These mating opportunities expand exponentially as males are drawn into cooperative breeding activities, where supplemental provisioning assistance to females may serve as a ticket to both sexual access and group membership.

Such economic participation need not be construed as the cornerstone of male philopatry, pair bonding, or sexual exclusivity. For males, who would leave their mother's kin group at adolescence, access to fertile females becomes equated with residence in non-natal groups. Assurance of paternity takes a back seat to the sexual access that a multifemale-multimale cooperative breeding community offers. By providing tangible benefits to females, males are integrated into the community, thereby increasing their reproductive success. It is at this point, where male and female inclusive fitness goals become aligned, that reproductive energy is optimized and that the genus *Homo* emerges.

Finally, this scenario proposes that Early Plio-Pleistocene humans subsisted not on big-game hunting, but on the intensive and opportunistic gathering of tubers, scavenged meat, and lipid-rich resources in the lacustrine food chain. It also recognizes the central role played by females in making and utilizing tools, expanding the range of high-quality foods to meet the species' growing nutritional needs, and developing critical food-sharing strategies for the provisioning of young.

While the scavenging of carnivore kills and hunting of small animals likely provided a supplemental source of dietary protein, it has been argued here that fish, shellfish, and other aquatic resources composed a central component of the early *H. erectus* diet. These foods, rich in amino and fatty acids, were abundant, easily harvested, and a source of essential nutrition for both brain expansion and population growth. Dependency of *H. erectus* on these resources would have had a determinative effect on population densities and movements within Africa, where climatic and tectonic events changed the flow of watercourses and alternatively filled and emptied paleo-lakes. Cyclical declines in aquatic flora and fauna initiated by processional dry periods may have also triggered the periodic movements of *H. erectus* into Eurasia. When resources became depleted or when the carrying capacity of habitats was exceeded, early human groups moved. And when they moved, they likely did so not to follow herds of mammalian herbivores, but to follow the wetland habitats familiar to them—the rivers and shorelines—the water and the bounty it offered.

This alternative scenario for the emergence of genus *Homo* represents a dramatic departure from traditional theory, which has been firmly grounded in the framework of game-hunting economies. Because of the important role assigned to Paleolithic diet in the trajectory of human evolution, chapter 4 takes us on a brief detour to re-examine the wider array of foods potentially exploited by ancient humans and how these opportunistic subsistence patterns influenced not only where they lived, but the fabric of their social lives.

 4

PALEOLITHIC DINNER PAIRINGS
Red or White?

Theories on human sapientization, though divergent, generally agree on three basic premises. The first is nutritional sufficiency. Evolution of the genus *Homo* was jump-started and sustained by high-quality foods that provided both carbohydrates and essential amino and fatty acids. Principal among these dietary shifts was systematic carnivory—the regular consumption of animal flesh. The second area of consensus revolves around energy efficiency and energy redistribution that, in some models, translates into altruism, cooperation, and the division of labor. Early humans leapfrogged ahead of their contemporaries by becoming eusocial. And third, the rapid dispersal of *H. erectus* lineages across tropical and temperate regions of Africa and Eurasia was catalyzed by climate instability and its impact on available resources.

But, as the saying goes, the devil is in the details. And the details on ancient lifeways, particularly for the first million and a half years of human existence, are slim. What we *do* know is that *H. erectus* populations successfully colonized the entire middle latitude belt from western Europe to southeastern Asia during the Early Pleistocene. They also provided the ancestral stock for subsequent *Homo* lineages that culminated, hundreds of thousands of years later, in Neanderthals and anatomically modern humans (AMHs). The sparse paleontological and archaeological record, however, has left much room for speculation on the nature of this evolutionary journey. This chapter pauses to explore how theories on the nature of hominin diets have influenced the reconstruction of Paleolithic social life.

The Two Faces of Carnivory

A common approach to filling data gaps in the fossil record has been to dial the human experience backwards. That is, if Late Pleistocene AMHs represent the pinnacle of morphological and behavioral evolution, then *H. erectus* and subsequent intermediate forms represent formative steps in a linear progression from simple to complex, from primitive to advanced. By this logic, if early AMHs were proficient hunters of mammalian herbivores, then ancient humans must have been similarly reliant on the procurement of meat, albeit with lesser degrees of success. The hunting hypothesis proceeds on this premise—that the hunting of terrestrial animals guided the course of human evolution, both at its inception and throughout the entire Pleistocene. While other subsistence activities are recognized as contributory, it is the procurement and consumption of mammalian meat, we are told, that grew our bodies and brains, and that defined our socioeconomic adaptations through time.

The legacy of the hunting hypothesis is both subtle and far-reaching. It frames our perceptions of not only how members of the genus *Homo* made a living, but where they lived (and did not live), how their cognitive abilities were honed, how their populations radiated in time and space, the size and social fabric of their communities, and even why their lineages prospered or went extinct. The fundamental flaw of this hypothesis is that it tells only *half* the story, and in two critical ways. First, it seeks to understand the whole of a species' development by assigning the procurement of animal flesh to only half of its members. Ancient society is defined largely by notions about what males were doing through time. At its best, it is an evolutionary saga in which men provide the main course and women the side dishes. At its worst, female hominins become invisible, encumbered with babes in arms or simply banished to the hearth. This picture does not square well with current primate and ethnographic evidence, nor with what can be reasonably inferred about the ancient roots and importance of female provisioning.

A second shortcoming is that it considers only half of the protein universe. For evolving humans, carnivory has always had two faces: the scavenging and hunting of terrestrial animals, and the consumption of aquatic fauna. While exploitation of freshwater and marine animal foods has been acknowledged as important for some Pleistocene populations, the broader evolutionary significance of these protein and lipid sources is overshadowed in current models by the presumed reliance of early humans on terrestrial mammal meat.

Such theories link the exodus of *H. erectus* from Africa to climate-driven changes affecting the distribution of mammalian prey. Increasing aridity and the expansion of grasslands served as a "Paleolithic pump," expelling savanna-adapted early humans from the continent in pursuit of water and the herd animals that sustained them. Their spread into eastern Asia, the Caucasus, and western Europe over the next million years has been similarly linked to the consumption of meat and the mastery of fire making (Boaz 1997). But, as noted in chapter 3, the reliance of *H. erectus* on a hunting subsistence economy over this span of time and space is not supported by the archaeological record.

Homo has always had a taste for meat and marrow. But dietary reliance by evolving humans on mammalian flesh as their primary protein source has varied greatly over the past 2 million years. Generally speaking, the efficiencies and potential returns of meat procurement efforts in the Early Pleistocene appear marginal. Activities were largely limited to the scavenging of carnivore kills, supplemented perhaps by the taking of small terrestrial animals in proximity to encampments. When viewed across the span of human evolution, the organized hunting of mammalian herbivores was slow to develop. In understanding why this is so, it is perhaps useful to consider the circumstances under which mammalian meat became a component of ancient diets—to start at the beginning and dial the human experience forward.

Successful procurement and consumption of meat by early hominins was constrained by both biology and technology. The most fundamental of these limitations was nutritional. Recent studies suggest that meat consumption may have been importantly impacted by threshold health issues. Smith et al. (2015) studied bacterial contamination levels in carrion typical of carnivore kills scavenged by early humans. Their study found that bacterial loads on raw carrion flesh rose to potentially dangerous levels within twenty-four hours.[1] Biological hazards also included parasites. Anton (2003: 60), along with Boaz and Ciochon (2004: 105–7), cite the results of genetic research conducted by parasitologist Eric Hoberg et al. (2001) that traced the evolutionary history of tapeworms. These studies found that the three primary mammalian tapeworm species had already colonized the human digestive tract by 1.7 ma, suggesting an earlier extended period of raw flesh ingestion by early hominins. In short, it is likely that food poisoning and tapeworm infestations were not uncommon among the initial consumers of animal meat. The source of these ailments, along with the remedy, may have been deduced independently and early on by *H. erectus* groups. In both instances, fire-roasting carrion and other meats could significantly reduce or destroy harmful bacteria and tapeworm larvae. The antiquity of fire making and its use for this purpose in Pleistocene Africa and Eurasia, however, is still an unresolved question. Heavy reliance on meat absent an appropriate balance with dietary fats has also been linked to calciuria and osteoporosis in Middle Paleolithic populations (Cachel 1997).

A second and pivotal factor in the dietary importance of meat is the advancement of weaponry. Prey animals must be safely and reliably taken in sufficient quantities to make a significant contribution to early hominin nutrition. The Mode 1 tool assemblages of early *H. erectus* are ideally suited to the collection of scavenged meat and marrow and the gathering and processing of aquatic fauna. But lithic projectiles are lacking. It is conceivable that fish and small animals or larger disabled prey could have been taken by hand or with sharpened sticks or crude choppers. But the routine killing of medium to large terrestrial mammals in the Late Pliocene-Early Pleistocene is not indicated. Indeed, prey animals, even when occupying the same habitat, may have been passed over as a staple food source due to the difficulty or danger in pursuing them. This is particularly true if alternative sources of protein, such as fish and shellfish, were nearby, abundant, and easily obtained.

Similarly, the Mode 2 assemblages that spread from Africa to much of Eurasia during the Early Pleistocene lack obvious weaponry. Functions of the signature Acheulean biface handaxe have been variously attributed to butchery of animal flesh, wood chopping, the scraping of hides, and root digging—but not as a tool of predation. The association of Acheulean assemblages with some large-animal kill sites, while initially thought to indicate hunting activity, has been later attributed to the persistent pattern of carnivore-kill scavenging. It is not until the emergence in Europe of *H. heidelbergensis* or proto-Neandertaloids in the early Middle Paleolithic period that evidence for systematic hunting of terrestrial animals appears (Conard et al. 2015).

The question then becomes, what else did early humans eat for the first million and a half years? What other sources of protein and fatty acids essential for growing our bodies and brains were readily available? It is proposed here that aquatic flora and fauna played a significant and complementary role in the emergence and subsequent evolution of the genus *Homo*. Evidence for regular exploitation of these resources in eastern Africa by Late Pliocene hominins is now documented. Given this heritage and the proficiencies of early humans for exploiting aquatic foods in areas of their geographical origin, it is reasonable to assume that they would continue to procure hydrophytic plants and littoral or freshwater fauna as they expanded throughout the Old World. Indeed, settlement in close proximity to shorelines and wetlands is the dominant land use pattern for archaic humans throughout the Pleistocene.

Nicholas (1998) examined the close association of both ancient and modern hunter-gatherers with wetlands such as swamps, estuaries, and marshes, noting the importance of these habitats as sources of diverse, abundant, and reliable foods. Nicholas comments that although this connection is clearly reflected in African and Asian fossil hominin sites, wetlands information is generally underrepresented or "invisible" in the archaeological literature. He concludes: "Without recognizing the importance of wetlands, we will not have a representative or complete view of the past" (1998: 728).

Finlayson (2004, 2009, 2014) also draws attention to the ubiquitous association of ancient human populations with rivers, lakes, floodplains, marshes, swamps, estuaries, and marine shorelines. This waterside settlement pattern prevailed across the entire breadth of Eurasian middle latitudes, from Spain to China. Living in proximity to water is linked to both human hydration requirements and the selection of habitats with high prey densities. Finlayson proposes that fresh water sources and coastal areas were favored because of their frequent association with forest margins and open spaces that supported a variety of mammalian herbivores. Notably, however, these are the same habitats that also offered an abundant and easily accessible supply of protein- and lipid-rich aquatic foods. Finlayson acknowledges the existence of coastal and wetland occupations by both *H. erectus* and Neanderthal populations, as well as habitat effects on diet such as the marine influence in Southeast Asia. But in the final analysis, aquatic resources clearly take a back seat to the scavenging and hunting of mammalian herbivores as the dominant food quest theme of ancient humans.

In his comprehensive examination of prehistoric aquatic adaptations, Erlandson (2001, 2010) maintains that reconstruction of Paleolithic subsistence and settlement patterns is hampered by both the incompleteness of the archeological record and persistent theoretical biases. The notion that hunting by males and meat consumption formed the cornerstone of ancient human existence remains popular in contemporary archaeology. Dubbed by Erlandson as the "hunting hangover," this paradigm has as its corollary an a priori disregard for protein sources such as shellfish, fish, waterfowl, sea birds, and amphibians at prehistoric sites and the systematic downplaying of their potential significance. Aquatic fauna are frequently lumped together as "small package" or "gatherable" foods, the collection of which is claimed to compare unfavorably with meat procurement in terms of energy expenditures versus returns. Foraged sources of protein have often been characterized as "backup" rather than primary resources, and as having lower productivity as a subsistence strategy than terrestrial hunting in terms of sustainable yields. Erlandson also cites studies which suggest that some aquatic resources, such as shellfish, are devalued as dietary staples simply because of their assumed association with female rather than male subsistence activity (Claassen 1991; Moss 1993).

Biases about the productivity and importance of aquatic resources ignore their widespread abundance and accessibility in the land use areas favored by Pleistocene hominins, such as the Mediterranean coast (Colonese 2011). Shorelines and wetlands provided sources of both marine and freshwater shellfish, as well as opportunities for the mass harvesting of spawning fish and of large schools of small fish native to shallow waters. Such habitats offered the potential for sizeable yields with minimal energy expenditures and without any specialized technology. Aquatic resources available to early humans also included larger protein packages such as marine mammals, which could have been taken with relatively simple tools when they were incapacitated or otherwise beached in birthing and breeding colonies.

Ideally, questions about the relative dietary importance of terrestrial and aquatic fauna are resolvable through an examination of the archaeological record. Erlandson (2001) provides examples of aquatic adaptations associated with *H. habilis, H. erectus,* and other archaic humans. The problem, however, is that sites from the Early Pleistocene period are both small in number and poorly documented. Erlandson suggests that evidence of aquatic diets at other ancient sites may be compromised by the differential preservation of fossil material, which favors higher density bones of terrestrial mammals over that of smaller littoral or freshwater fauna, such as fish. He also notes that the interest and field methodology of the investigator may skew archaeological site findings, which historically have focused on the recovery of large land animal bones to the exclusion of other skeletal faunal remains. Additionally, bones of terrestrial mammals are frequently associated with butchery rather than habitation sites and therefore present only a partial picture of early human diets.

A potential consequence of the hunting hangover is the failure to consider sources of dietary protein other than terrestrial mammal meat, such as aquatic fauna, that may be nearby and readily available. For example, Boaz and Ciochon (2004: 154, 174), in their examination of Early Paleolithic occupations in Zhoukoudian, China, and Sundaland, Southeast Asia, concluded that scavenged meat was the source of *H. erectus*'s high-calorie, high-protein diets, but admitted that the range and seasonal procurement of terrestrial prey animals remains a "mystery." No discussion is given to the potential importance of aquatic fauna in these ecological settings, despite the proximity of the Zhoukoudian cave site to a river with abundant ancient fish populations, or to other aquatic fauna in the resource-rich corridor that dominated the riverine, estuarial, deltaic, and southern coastal regions of the Southeast Asian peninsula.

Perhaps the greatest impediment to reconstructing Early Pleistocene lifeways was created by glacial effects on the paleogeography of the African and Eurasian continents. Virtually all coastlines existing one hundred twenty thousand to fifteen thousand years ago are now submerged and distant from the current shoreline. Consequently, the archaeological record for early hominin coastal occupations has been largely obliterated through flooding, erosion, and ocean inundation. This raises the perennial dilemma referenced by Kuhn and Stiner (2006: 970), namely, how to distinguish the "absence of evidence" from the "evidence of absence."

The Question of Dietary Breadth

Since, predictably, the bulk of known prehistoric coastal sites are dated after the Last Glacial Maximum, several scholars have proposed that aquatic adaptations are not ancient at all, but rather are a relatively recent phenomenon associated primarily with AMHs. Those ascribing to this view concur with the premise that hunting of mammalian herbivores was the dominant subsistence strategy of ancient humans until the Terminal Pleistocene/Early Holocene. They are therefore left with the task of explaining why and where these late-period aquatic dietary shifts occurred.

One explanation grew out of a cross-cultural study of housing and settlement patterns among contemporary hunter-gatherers in northerly latitudes. Binford's (1990) "aquatic revolution theory" proposed that initial postglacial occupants of polar margin areas were forced to expand their dietary breadth from terrestrial mammals to aquatic fauna due to the lack of local plant material for cold weather stores. Aquatic resources, he argued, were targeted as foods for storage, a strategy for reducing group mobility during the winter months. Accordingly, he saw the aquatic revolution as occurring first in northern latitudes and proceeding southward over time during the Terminal Pleistocene. Holliday (1998) concurred with Binford that a shift toward reliance on aquatic resources originated in northern latitudes. He pointed to postglacial reforestation of the tundra, the extinction of megafauna, and the disappearance of gregarious prey species as principal factors in the dietary transition of AMHs to fishing and trapping in the Early Holocene period.

Cachel (1997) also proposed that key dietary shifts occurred in the northern latitudes of Europe. She argues that Neanderthals were susceptible to developing calcicuria and osteoporosis, conditions associated with protein/dietary fat imbalances that arose from heavy reliance on meat and the scarcity of edible plants. Critical protein/fat ratios were allegedly improved by the adoption of more efficient hunting techniques and consumption of fat-rich prey animals. Cachel maintains that there is no evidence for the reliance of Neanderthals on marine resources or anadromous fish (1997: 583). It is not until the Late Upper Paleolithic that the protein/fat ratio issue is ultimately resolved, presumably by AMHs, through more intensive hunting of mammalian herbivores and the expansion of dietary breadth to include fats and oils from both vegetative and aquatic sources.

Finally Stiner (2002), along with Kuhn (Kuhn and Stiner 2006), join the others in assuming that the *H. erectus* protein diet was limited to scavenged meat. Stiner opines that early hominins in Eurasia did not penetrate areas above 50^0 in latitude prior to 500 ka, presumably constrained by glacial conditions and the presence or absence of fire making. This was followed by three major "niche boundary shifts" at 500 ka, 250 ka, and 50 ka, associated, respectively, with prime-adult-biased large-animal predation; the expansion of foraging substrates to include small animals in response to increasing population densities; and finally by dramatic socioeconomic and energy efficiencies enabled by advanced technology that coincided in the Late Upper Paleolithic with the appearance of AMHs.

Kuhn and Stiner (2006) acknowledge the existence of Neanderthal aquatic adaptations in the lower latitudes of the Mediterranean Basin during the Middle Paleolithic period. But they are quick to dismiss these occupations as outliers that were precipitated by increasing population density and resultant resource stress. Coastal sites are seen as reflecting a shift in subsistence strategies to smaller game, which included aquatic "gatherables." While admitting that the exploitation of shellfish and small animals at such sites could result in net yields similar to those of medium-to-large-sized game, Kuhn and Stiner argue that such resources were easily overexploited and unsustainable due to the slow reproductive recovery rates of aquatic fauna. It is only after 50 ka, they propose, and particularly with the mastery of watercraft and exploitation of free-swimming fish after 20 ka, that coastal occupations by AMHs became viable.

In summary, theorists who discount the antiquity of aquatic adaptations draw parallel conclusions about the subsistence strategies of early humans prior to the appearance of AMHs. A common thread is the assumption that their diets were narrowly focused on terrestrial mammal meat, so much so that, in some instances, this dependency led to health impairment. The selective addition of aquatic fauna to Pleistocene diets is variously explained as a deviation from or supplement to preferred mammalian meats. In each case, this shift is pursued as a solution to a perceived problem. Thus, exploitation of new aquatic resources is seen as allowing northern latitude groups to secure stores that permitted greater sedentism during winter months; to compensate for meat shortages related to the climate-induced extinction or migration of large game animals; or to augment dietary transitions to small game caused by population pressure and resource stress. In all of these scenarios, AMHs ultimately emerge as the superior problem solvers, expanding their dietary breadth and economic diversity to eventually exert dominion over indigenous populations by the end of the Pleistocene.

Theories that question the antiquity of human aquatic adaptations are not without challenge. Critics point to the selective use of supporting archaeological data as well as the narrow geographical focus of these scenarios on the northern latitudes of Terminal Pleistocene Europe. Finlayson (2009: 88–89) discounts the idea that coastal and aquatic resources were discovered relatively recently by ancient humans. He notes evidence for early coastal occupations at a formerly submerged reef terrace on the Red Sea coast of Eritrea dated 125 ka (Bruggemann et al. 2004), and at the 165 ka Pinnacle Point cave site in South Africa, which was spared sea inundation by its elevation (Marean et al. 2007). Finlayson also cites more ancient examples of *H. erectus* coastal and shoreline occupations such as those at Trinil, Java, dated 1.5 ma (Joordens et al. 2009), on the coastal plain of India dated 1.5 ma (Pappu 2011), and at Ubeidiye, the Jordan Valley, dated 1.4 ma (Belmaker et al. 2002).

Because aquatic revolution scenarios focus on the Late Upper Paleolithic/Early Holocene period, much of the debate on Pleistocene diet has centered on the geographical distribution and subsistence patterns of Neanderthals. Neanderthals are commonly assumed to have occupied high latitude habitats and to have relied on a narrow-spectrum mammalian meat diet due to the limited availability of plant resources. Their alleged lack of dietary breadth has been contrasted with the more broad-spectrum repertoire of AMHs and utilized to argue against the earlier exploitation of aquatic foods. Shea (2006) cautioned against drawing generalizations about Neanderthal subsistence patterns based on a limited number of comparatively recent European sites, noting their broad geoecological range. Soffer (2006) also questioned whether the archaeological record supports the Middle and Upper Paleolithic subsistence shifts proposed by Kuhn and Stiner, arguing that foods consumed by Neanderthals were not dissimilar to those of AMHs in terms of dietary diversity.

The case for low-diversity Neanderthal diets is based on the alleged scarcity of carbohydrate-rich plants or tubers in northern latitudes, as well as the absence of food-processing artifacts such as grinding stones. Recent studies by Henry, Brooks, and Piperno (2011, 2014), however, found direct evidence for not only the consumption, but the *cooking* of plant foods by Neanderthals. Starch grains recovered from the dental calculus of Neanderthal fossils at Middle Pleistocene sites in Belgium and Iraq indicate that the occupants of both warm- and cold-climate habitats were gathering, processing, and eating a variety of local plant foods. Plants linked to the Shanidar Cave site in Iraq included grains associated with the wild relatives of wheat, barley and rye, and phytoliths from the date palm. And, significantly, at the northern latitude Spy Cave site in Belgium recovered microfossils included starches from plants aligned with the submerged underground storage organs (USOs) of water lilies. These findings indicate that Neanderthals living in diverse climate zones enjoyed diets of considerable breadth and, moreover, knew how to transform edible plants into more digestible and nutritious foodstuffs by cooking. They also suggest that Neanderthals occupying northern Europe were *in the water* gathering plant foods, and hence perhaps also aquatic fauna, well in advance of the Terminal Pleistocene.

Finlayson (2004, 2009) has effectively challenged many of the long-held myths surrounding the evolutionary history of Neanderthals, including where they lived and how they made their living. Prominent among these myths is that Neanderthals were a highly specialized lineage—some say species—that evolved and thrived in the frigid northern latitudes. The notion that their stocky body morphology was an adaptation to cold temperatures has become a textbook legend, with some scholars even opining that Neanderthals had a low tolerance for warmer climes (Boaz 1997: 210–12). Finlayson's findings have turned this model on its head. He concludes instead that Neanderthals evolved in the Eurasian middle latitude belt, expanding during their heyday in an east-west direction from Portugal to Siberia. They ventured onto the northern European plains during milder Pleistocene events, where the hunting of large mammalian herbivores assumed a prominent role. But Neanderthals also migrated south with each glaciation, occupying the warm and humid refugia of the Mediterranean peninsulas, the Caucasus, and Crimea, where coastal settlement was not uncommon and where aquatic resources, such as shellfish, seals, fish, and cetaceans, were regularly exploited. Neanderthals occupying regions such as the floodplain of Italy's Po River had access to a wide range of aquatic fauna, including the annual runs of *Esox lucius*. As portrayed in Figures 4.1 and 4.2, these large Northern pike could be taken in shallow

Paleolithic Dinner Pairings • 107

Figure 4.1. "Lucius." By Emiliano Troco, oil on canvas, scientific supervisor Davide Persico, private collection.

Figure 4.2. "Neanderthal Clan." By Emiliano Troco, oil on canvas, scientific supervisor Davide Persico, collection of Museo Paleoanthropologico de Po.

waters close to the river bank where they came to spawn each spring, and reportedly reached lengths of nearly six feet and weights exceeding sixty pounds.

The Gorham's Cave site in Gibraltar provides another example of dietary breadth. Neanderthals occupied the site commencing around 125 ka and left evidence of a consistently broad foraging regimen spanning the next hundred thousand years. These peoples consumed a wide variety of resources including ibex, limpets, mussels, birds, rabbits, tortoises, monk seals, fish, dolphin, and highly nutritious stone pine seeds (Finlayson 2004: 98–99). A similar reliance on marine resources is noted for contemporaneous hominins living in South Africa. Finlayson concludes that the consumption of a broad spectrum of prey animals is probably a very ancient pattern among humans, who optimized their foraging strategies in response to resource availabilities and climate challenges in a wide variety of spatiotemporal environments.

This interpretation of Neanderthal origins and lifeways is significant in that it points to their geographic and phylogenetic continuity with earlier human populations in the Eurasian middle latitude belt. This same region of the Old World was home to the pre- and early-*H. erectus* lineages that pulsed out of Africa during the Late Pliocene and Early Pleistocene, first via the Near East and later via Gibraltar and the Iberian Peninsula.[2] The middle latitudes are also home to a number of subsequent hominin lineages that evolved in situ. Some of these went extinct, but still others provided the foundation for more advanced humans. For example, an early hominin fossil from Ceprano, Italy, dated 800–900 ka, displays morphological characteristics intermediate between *H. erectus/ergaster* and *H. heidelbergensis* (Manzi, Mallegni, and Ascenzi 2001). This and a contemporaneous fossil from Atacuerpa, Spain, have been classified as early forms of *H. heidelbergensis,* or alternatively as a new species, *H. antecessor* (Carbonell et al. 1995). Later fossils dated 430 ka from the Sima de los Huesos site in Atapuerca have been more directly linked to proto-Neanderthals (Arsuaga et al. 2014).

In addition to sharing a phylogenetic connection, ancient humans occupying the Eurasian middle latitude belt shared the same lithic technology for about 1.5 million years and thus, arguably, a similar range of subsistence strategies. It has been proposed here that *H. erectus* populations left their homeland already equipped with the cognitive abilities, knowledge, and skills to pursue a broad-spectrum diet based on the gathering of edible plants and the procurement of both terrestrial and aquatic fauna. Just as mammalian meat was a mainstay, so too were myriad sources of protein- and lipid-rich aquatic fauna. The ability to opportunistically pursue multiple sources of animal protein, to anticipate and locate these resources in time and space, and to tailor foraging strategies around their availability was a hardwired trait of *H. erectus* and all subsequent human lineages. The focus and relative breadth of hominin diets is a reflection of the changing inventory of resources at hand and the ability to efficiently procure them.

The wildcard in this evolutionary drama has been the cyclical effects of climate change on the distribution of faunal resources. Finlayson (2004) has reconstructed a portrait of mammalian herbivore prey distributions throughout the Pleistocene in relation to climate events and variable vegetation structures. This has facilitated an understanding of how changing habitats and the range shifts of terrestrial animals may have affected the geographical distributions and subsistence patterns of early human populations through time. But in order to complete the picture of protein resources available to Pleistocene populations and the influence of climate on their distribution, the same type of information must be assembled for coastal and inland water habitats and for the marine and freshwater fauna they supported. To this writer's knowledge, a comprehensive analysis of how glacial cycles affected the distribution, viability, and accessibility of aquatic resources in these habitats has not been completed. Understanding the climate change effects on this *other half* of the dietary protein universe is equally important because it also impacted the distribution and diets of ancient humans throughout the Pleistocene.

Documentation of these effects is the domain of marine biogeochemistry and lies beyond the scope of this book. But suffice it to say that glacial advances and retreats affected the migratory patterns of marine mammals, cetaceans, fish, birds, and other aquatic fauna. The dramatic temperature, sea-level, and seawater chemical changes accompanying glacial events are associated with both the extinction and reradiation of marine fauna. Stressful or even fatal environments for plankton, oyster, and clam species have been linked to seawater salinity changes accompanying glacial melt, while the viability of coral reefs, along with the life forms they support, is known to be negatively impacted by sea level changes. Glacial events therefore have the potential to decrease the availability of both marine nutrients and aquatic fauna in proximity to shorelines during certain glacial phases and then to restore these resources and the productivity of coastal occupations at the other end of glacial cycles.

For example, lowered sea levels are known to have disconnected Mousterian cave sites in Italy from the shoreline, potentially reducing local dependence on marine mollusks (Kuhn 1995). The effects of changing sea levels on marine resources and on human settlement patterns have also been noted for the Northwest Coast Indians, who are thought to have arrived in this region of North America before 13 ka. Layton (2008) notes that rising sea levels accompanying glacial melt disrupted coastal and estuarial fauna, including the viability of shellfish beds and riverine spawning runs. It was only after the stabilization of sea levels and recolonization by marine life after around 3500 BC that permanent coastal Indian settlements were established. Such changing ecological conditions may account for the cyclical occupation and abandonment of coastal sites throughout the Pleistocene. In this important sense, the resurgence of coastal adaptations after the Last Glacial Maximum may not represent a broad-spectrum revolution at all, but simply the continuation of an ancient pattern of accommodation to climate effects on marine resource availability.

A similar investigation is needed on the impact of glacial advances and retreats on freshwater aquatic fauna throughout the Pleistocene. Certainly in central Africa, fishing with bone harpoons was undertaken by Middle Stone Age peoples well in advance of the Terminal Pleistocene in Europe (Yellen et al. 1995). Indeed, based on Late Pliocene evidence of Oldowan tools and abundant ancient fish remains in the upper Semliki River Valley region (Boaz 1990), reliance on freshwater fish may well span the entirety of hominin existence. Proponents of aquatic or broad-spectrum revolution theories tend to write off such examples as exceptional, insisting that reliance on freshwater aquatic fauna, particularly in Europe, was a late development. But again, the absence of evidence is not the same as evidence of absence. Documentation of the spatiotemporal distribution and potential availability of waterfowl, the European catfish, Northern pike, and of anadromous species such as Atlantic salmon, allis shad, and sea lampreys during earlier periods of the Pleistocene is critical for at least two reasons. First, early inhabitants of Europe lived in proximity to lakeside and riverine habitats compatible with these species. They presumably would not have been blind to the value of these fauna as food sources. Second, dietary reliance on such abundant, reliable resources has significant implications for early human settlement patterns, population density, and social life.

Labor Division, Productivity, and Complexity

Before leaving the question of early hominin diets, there is another dimension of Pleistocene economies that must be addressed, namely, how our ancestors organized their communities and allocated work to meet the challenges of changing Ice Age environments and landscapes.

Recently, archaeological theories have expanded their domain from descriptive interpretations of material culture to speculation on the broader socioeconomic characteristics of Pleistocene populations. A focal point of this discussion has been the surge in complexity witnessed in the Late Upper Paleolithic period and its seeming nexus with the arrival of AMHs in Eurasia. To what can dramatic increases in productivity, population density, sedentism, and social networks be attributed?

Kuhn and Stiner (2006) have proposed that the key to adaptive success for AMHs lies not only in what they ate, but how they organized their economic activities. They argue that immigrant populations prevailed over contemporaneous Neanderthals by shifting to a reliance on more varied small-animal resources such as birds, quick-flight fauna, and eventually fish. This dietary broadening, which also included supplementary vegetable foods, involved the exploitation of multiple resource types that were often separated by location and by the seasons. Such diversified subsistence strategies, they propose, selected for labor specialization on the basis of age, gender, and ability. By moving to a lower trophic level, and by organizing their economies around cooperative and complementary subsistence roles, AMHs were able to lower dietary risks, enhance productivity, increase population density, and realize larger and more sedentary communities. Kuhn and Stiner theorize further that the division of labor by sex was more likely to evolve in the tropics and subtropics due to the greater biotic diversity such areas offered, and hence greater opportunities for expansion of dietary breadth and foraging specialization. Since AMHs are thought to have originated in low-latitude regions, it is suggested that they may have already been preadapted to pursue what are referred to as "collaborative economies" prior to their dispersal from Africa.

Kuhn and Stiner contrast this AMH scenario with what they perceive as the more monotypic and demographically limiting economy of Neanderthals based on the hunting of large terrestrial mammals. Due to the challenges of subduing and killing large prey without sophisticated projectile weaponry, they theorize that Neanderthal subsistence activities required the communal efforts of all group members—men, women, and children—either for direct confrontation with prey animals or for driving them into ambush. The heavy reliance on meat, when coupled with occupancy of plant-deficient habitats, militated against the exploitation of diversified resources and the development of distinct and complementary subsistence roles. Consequently, Kuhn and Stiner propose, Neanderthals lacked a division of labor by sex. Absent the greater productivity that a broader resource base and economic specialization could bring, Neanderthals are perceived as occupying a kind of demographic dead end, doomed to life in small, mobile communities with narrow, high-risk diets and limited potential for population growth. In the end, this leads to their undoing.

While this theory presents some interesting contrasts, it tends to unravel when the range of Neanderthal subsistence economies is more fully explored. Its basic premise rests, in large part, on the generalization that Neanderthals lacked a broad-spectrum diet, without which "collaborative economies" would fail to develop. As noted earlier, such characterizations portray Neanderthals as a cold-adapted species that inhabited sparsely vegetated northern latitudes on a permanent basis. However, evidence now suggests that such occupations were largely limited to warm interglacials and that even peoples that hunted on the northern plains during these periods included plant resources in their diet. As Finlayson has convincingly argued, Neanderthals originated and spent much of their time in the temperate middle latitude belt where habitats, resources, and subsistence economies were more varied. Kuhn and Stiner write off as short-lived or unsustainable Neanderthal Middle Paleolithic sites in the Mediterranean Basin that contain evidence of small-animal and aquatic resource exploitation. But other signature sites are more difficult to dismiss in terms of their model. Significantly, over 80 percent of the skeletal remains of animals consumed by Neanderthals at the Gorham Cave site in Gibraltar belonged to rabbits, while large mammals constituted only a small portion of their diets (Finlayson 2009: 149–53). Similarly, Neanderthals occupying the temperate river valleys of the European continent that emptied to the sea had a varied menu of protein sources from which to choose.

What this tells us is that broad-spectrum diets were not unique to AMHs. Some Neanderthal populations were already pursuing similar subsistence strategies for millennia prior to the arrival of AMHs in Europe. It also suggests that related attributes such as labor specialization, enhanced productivity, population density increases, and greater sedentism may be equally applicable to Neanderthal populations and perhaps also to earlier lineages that were able to carve out successful food-getting strategies in biodiverse habitats. As illustrated by Early Upper Paleolithic sites in the Levant, dietary breadth and the organization of work among early human groups was highly plastic, fluctuating with the seasons and with the broader climatic trends affecting resource availability (Binford 1968). The distribution of flora and fauna, technology, and the carrying capacity of a given niche are the critical drivers of ancient land use and labor division patterns. In this important sense, the socioeconomic structure of Neanderthal and AMH communities is more a reflection of ecology than phylogeny.

Toward More Inclusive Models

Remedies suggested by Erlandson (2001) for reconstructing a more accurate and complete picture of Pleistocene subsistence patterns include renewed archaeological excavations and more sophisticated analytical techniques at early sites, along with examination of submerged landscapes for evidence of shell middens and other indicators of coastal settlement. While future archaeological sites may shed additional light on the fluctuating importance of terrestrial and aquatic protein in early hominin diets, the fact that much of the evidence has been overcome by geological events will continue to cloud the picture of Pleistocene adaptations. Erlandson notes:

> Until anthropology transcends some pervasive misconceptions, the significance of aquatic adaptations will continue to be underemphasized in our reconstructions of human evolution. These misconceptions include (1) the notion that large land mammals were virtually always the most productive and highly ranked resources for our hominin ancestors; (2) that male-dominated hunting was always the central force that shaped human subsistence, settlement, and technological developments; (3) that the utilization of aquatic resources is automatically evidence for demographic pressure or resource stress; and (4) that the archaeological record preserves a representative picture of our past. (2001: 305)

This chapter's title posed a metaphorical question about Pleistocene diets. A reasonable answer is that if early humans were sommeliers, they would have probably kept red *and* white varietals in their wine cellars. When it comes to food, genus *Homo* was an opportunist, living near and taking full advantage of the full range of protein that the animal kingdom had to offer. Documenting the journey of how we evolved as an omnivorous species has been facilitated in recent years by a growing fossil inventory and a greater understanding of how Pleistocene geological and climatic events have shaped the biocultural trajectory of our ancestors. The paleontological and archaeological records available for reconstructing the past, however, remain woefully incomplete. Theorists will therefore continue to rely on speculation to fill evidentiary gaps.

It's been suggested throughout this discussion that new scientific data are calling some popular theories into question or are rendering them obsolete. It's also been noted that while evolutionary theories may be couched in the language of scientific discourse, they are not immune to cultural bias. Monotypic portraits of Paleolithic life are being challenged by new conceptual models that address the potential for greater diversity in early human subsistence pursuits, social forms, and community organization. Going forward, we are obliged to renew and refine our recovery efforts for ancient remains, rethink conventional wisdoms, and pursue the enlightenment that new cross-disciplinary research may bring to bear on how natural selection and climate change influenced the way our ancestors lived.

 5

Signature Hominin Traits

Reconstructions of human evolution typically begin by establishing general categories of fossil hominins and then proceed to arrange them and associated stone tool traditions along a chronological gradient of biological and cultural advancement. Our precursors are thus often represented as a set of standard deviations from modern humans, who occupy the pinnacle of evolutionary development. Such schemes, by definition, focus on the perceived *differences* between hominin lineages as a function of their phylogenetic rank. The flip side of this approach is to look instead at their *similarities*—at traits that characterize the common hominin experience through time. What are the essential qualities that define and bind together ancient and modern humans? These shared trait complexes are the product of mosaic evolution, the components of which emerged at different times and at different rates across a myriad of hominin lineages in response to selective pressures. Identifying these commonalities helps to highlight the evolutionary interface among biology, ecology, and natural selection. It also serves to remind us that we may not be so different after all. This chapter discusses seven of these signature traits.

Opportunistic Omnivory

There is an old story that maintains that if a banquet table was set up with a wide variety of nutritious and tempting dishes, and if a child was then provided free access to this feast with no restrictions on food choices, he or she may initially gravitate toward the confections, but given time would independently make broader and more balanced selections from all healthy food groups. The story's implication is that there is a little bit of genetic memory, some "nutritional wisdom," in all of us that tells us that dietary breadth is good. Whether or not there is any scientific merit to this anecdote, omnivory remains a defining characteristic of the genus *Homo*. From the beginning, hominins appear to have been opportunistic in what they ate, driven toward the optimal procurement of available plant and animal resources in a given habitat.

Initial dietary expansions were occurring by the Late Pliocene with new adaptations to terrestrial life. It is proposed here that the foraging strategies of female provisioners shifted to include not only a wider range of plant foods for themselves and their offspring, but an array of protein sources such as bird's eggs, small terrestrial animals, and aquatic fauna from local water sources. Scavenged mammal meat was gradually added to this cuisine and remained a supplementary food source for millennia prior to and after the development of organized hunting. The land use preference of ancient hominins for coastal, lakeshore, riverine, and other wetlands locations aligns closely with habitat types most likely to support biodiversity, and hence a wide range of foods. Omnivorous diets, and in particular diets including animal flesh high in essential fatty acids (DHA and LC-PUFA), are linked to ongoing hominin encephalization in the Early and Middle Pleistocene.

As noted in the previous chapter, there is no evidence to suggest that broad-spectrum diets were an invention of Late Upper Paleolithic peoples. Indeed, the distinctive characteristics of hominin dentition speak to the dietary diversity that came with early omnivory (Ungar 2012, 2017). Both *H. erectus* and Neanderthal populations are known to have occupied biodiverse habitats where they subsisted on a wide range of terrestrial and aquatic flora and fauna. Habitats varied both seasonally and over time in the variety and abundance of resources they offered, but recent evidence suggests that even in extreme latitudes early humans managed to procure an array of diversified foods.

The nutritional advantages of dietary breadth were greatly enhanced by a distinctively hominin invention: fire making and its use for food preparation and processing. Wrangham (2009) presents a convincing argument for the importance of cooking as a means of improving the digestibility and caloric value of foods early in human history.[1] Incremental increases in the energy budgets of *H. habilis, H. erectus,* and subsequent lineages attributed to cooked foods provided the foundation for changes in morphology (gut and dentition), better maternal nutrition, greater fecundity, extended life histories, and the fueling of bigger and more complex brains.

The subsistence activity of hominins is also marked by corporate food getting and food sharing, which serve as both a hedge against scarcity and the glue for social networks. Primeval food-sharing groups were founded on the altruistic relationship among mothers, offspring, and siblings. These social networks were likely expanded to include wider circles of individuals, such as alloparents, their kinsmen, and cooperative consorts, who forged economic relationships based on reciprocal benefit. There seems to be little basis for assuming, as does

Wrangham, that the nutritional leap forward initiated by the cooking of raw foods dictated new food-sharing patterns, i.e., that the invention of fire transformed the Pleistocene into a pair-bonded "Ozzie and Harriet" world based on the domestic servitude of women. Exactly who barbequed the antelope ribs, sea lion filets, and catfish, or fire-roasted the mussels and tubers, is unknown. What is likely, however, is that the members of hominin groups prospered by joining hands in both the quest for and partaking of food. These cooperative socioeconomic networks inevitably involved kin, and may or may not have recruited sexual partners on an ad hoc basis.

In summary, hominins evolved by maximizing their energy inputs from the broadest spectrum of available resources to meet the rigors of Paleolithic life. Contrary to the portrait of ragtag bands living from hand to mouth, many observers have suggested that the habitats occupied by our ancestors were uncrowded and generally replete with readily accessible foods that provided a nutritious, well-balanced diet. Indeed, it has been claimed that Paleolithic peoples were probably healthier, more physically robust, and longer-lived than many of the Neolithic populations that succeeded them.[2] Climate-related changes throughout the Ice Age impacted resource distributions and availability and necessitated periodic animal and human population movements. During extreme events, food and water shortages in certain areas undoubtedly created bottlenecks, resulting in population fragmentation and extinctions. But still other regions provided refugia where a variety of species congregated and were sustained. Until such time as more is known about the changing paleodemography of both terrestrial and aquatic species in the Pleistocene, a more detailed picture of how and where our ancestors lived throughout this era will remain incomplete. Hominins that lived to tell the tale and pass on their genes to the lineages that followed were what Finlayson (2009: 206–20) called "children of chance"—the risk takers whose behavioral flexibility, capacity for innovation, and life on the margins allowed them to respond quickly to environmental challenges.

Spatiotemporal Roadmapping

The successful conquest of terrestrial habitats by early hominins cast them into more complex relationships, both with their environment and with one another. The pioneers of this movement were challenged to broaden their perceptual field and make sense out of the diversity, character, and rhythms of novel biomes. New ecological adaptations emerged, affecting the spatial relationships of individuals and groups on the landscape, along with the architecture of their social networks.

Spatiotemporal roadmapping refers to the ability to perceive and conceptualize the external world, its resources, and its life forms in multidimensional ways. Acquiring the material necessities of life and functioning successfully vis-à-vis other group members requires individuals to create cognitive algebras capable of charting people, places, and things in geographical and social space. Moreover, these perceptual fields must be dynamic, reflecting past, present, and potential future states. Problem-solving experiments with a variety of contemporary primates, other mammals, and even birds suggest that such abilities are not unique to humans (de Waal 2016). These cognitive functions, however, are particularly elaborated among great apes and humans, suggesting that they are rooted in our *Pan-Homo* ancestral heritage.

For ancient hominins, survival depended on knowing both where food resources were located and when and under what circumstances they were likely to be found. This requires the mental capacity to not only imprint geographical information on resource locations within a home range, but to recognize and anticipate the life-cycle rhythms, natural processes, or other environmental conditions affecting their availability in the future. These aptitudes would have served our ancestors well as they abandoned arboreal adaptations for the more protean biospheres of terrestrial landscapes. The mosaic habitats in which our genus evolved contained a much broader range of potential floral and faunal food sources, the successful exploitation of which challenged them to become naturalists, observant of and attuned to the behaviors of all life forms around them. It also required an increasing awareness and anticipation of the circadian rhythms, celestial cycles, and seasonal changes that impacted the availability of local plant and animal foods, such as the flowering of plants; the ripening of fruits, nuts, and berries; the spawning runs of marine and freshwater fish; or the seasonal migrations of terrestrial herbivores, waterfowl, and sea birds. Armed with a knowledge of local taxa, along with the rhythms affecting their availability, hominins were able to realize greater efficiency in their food-getting activities and achieve the benefits of a broad-spectrum diet.

Spatiotemporal roadmapping was also critical for the establishment and maintenance of social networks. Among primates, the tracking of social space is central to group life, and involves at least three overlapping dimensions. The horizontal dimension establishes how individuals are distributed on the ground in relation to resources, kin and nonkin, and to external groups. As we have seen, ecological factors play an important role in the social demographics of primate groups, including mating patterns and the prevailing mode of philopatry. Social roadmapping provides information to individuals about which group members can be relied on for assistance and reciprocity and calibrates the reinforcement of strategic bonds through various forms of "social grooming." The vertical dimension refers to hierarchical relationships and establishes one's position relative to others in the exercise of power and status. Hierarchies serve an important regulatory function in most primate groups, structuring both reproductive and economic relationships. Successful group membership requires individuals to accurately map their place in the political landscape and to assess when to acquiesce to, collaborate with, or challenge extant power relationships. And finally, primate sociality often incorporates a temporal dimension, wherein social networks take on intergenerational depth. For example, individuals may be socially bonded by virtue of lineal descent from a common ancestor, living or dead.

Spatiotemporal roadmapping abilities are shared with nonhuman primates. Monkeys and apes know how and where to access appropriate foods. Many have demonstrated the capacity for innovation and planning, as in the gathering of materials at one location for use as tools at another time or place. All rely on social skills for survival. Social inequality is a fact of life for most primates, who must learn to cultivate social relationships that secure or advance their position within existing dominance hierarchies. Dominance is sometimes vested in specific lineages, the members of which are recognized as such by others, and whose economic and reproductive privileges may be intergenerational in scope. The scale of social networks among some primates may even extend beyond the local group to incorporate modular or multilevel communities. And the list goes on.

Despite being on separate evolutionary paths for the past several million years, the thumbprint of our common *Pan-Homo* ancestry remains. Metaphorically speaking, all primates could be characterized as geographic and social cartographers receiving, evaluating, and plotting information essential to their survival and successful reproduction in a given niche. When viewed side by side, the cognitive maps of hominins would be distinguished, perhaps, only by the number of data points and density of pathways, interchanges, and information networks. As the emerging field of evolutionary cognition reminds us, human abilities differ more by degree than kind.

For example, a three-dimensional picture of floral and faunal distributions provided early hominins information vital to the alignment of group size with seasonal or stochastic fluctuations in resource density. The pattern of alternatively dispersing and aggregating group members in cadence with available resources is shared with several contemporary primates. Hominins are distinctive, however, in the degree to which such fission-fusion patterns were formalized into expanded socioeconomic networks. High resource density areas or "hot spots," such as the seasonal migration routes of prey animals or rivers with anadromous fish runs, were favored throughout the Pleistocene as gathering places, often serving as the linchpin for modal or multilineage communities. Such locations were likely endowed with special significance through acts of communal subsistence activity and feasting, material exchanges, mating, and rituals that extended social networking into the symbolic domain (Gamble 1998). In other words, the social ties that bind were conceptually expanded beyond local face-to-face interactions to materials, artifacts, and landscapes that symbolized a larger sense of unity or belonging and that transcended local networks in both time and space.

What has been claimed as unique to our genus is the ability of individuals to recognize and participate in ever-widening social networks in the absence of physical copresence. It is with this "release from proximity" that hominins are proposed to have entered the realm of symbolic communication or "culture" (Rodseth et al. 1991). The ability of prehistoric peoples to maintain complex social ties among groups dispersed over large geographical ranges is frequently linked to the emergence of cognitive shortcuts—taxonomic structures or algebras (i.e., classificatory kinship terminologies, status and role categories) that distinguish and name the essential components of social networks and define their reciprocal relationships. The origin of language is typically credited with the unique ability of humans to coalesce family members, kinship groups, strangers, and even individuals who may never meet into increasingly complex, multilevel socioeconomic communities (Gowlett, Gamble, and Dunbar 2012). As noted later in this chapter, at what point this occurred in the course of human evolution remains a subject of debate.

Polygynandrous Mating

Polygynandry is a mating system, common to multimale-multifemale social groups, in which both sexes have access to multiple partners. It is the prevailing pattern among primates generally, including our closest living relatives (chimpanzees and bonobos), and is presumed to have characterized terrestrial Pliocene apes as well. Indeed, before much was known about paleontology or nonhuman primates in the wild, nineteenth-century evolutionists proposed this type of mating system, often referred to as "group marriage," as the earliest stage of human social life. The concept of primordial polyamorous unions, however, was largely swept aside by the antievolutionist fervor of early twentieth-century American and British anthropology. In its place emerged the idea that early hominins diverged from other primates by inventing sexual exclusivity (at least for females) and by turning casual or communal consortships into long-term reciprocal contracts. The male-female pair bond and resultant nuclear family thus became ensconced in popular theory as the cornerstone of human origins (Chapais 2008).

George Peter Murdock (1949) dedicated the entire first chapter of his classic work *Social Structure* to the virtues of the nuclear family. The man-woman-offspring triad was declared a universal solution for structuring the sexual, economic, reproductive, and educational functions regarded as basic to human social life: "This universal social structure, produced through cultural evolution in every human society as presumably the only feasible adjustment to a series of basic needs..." (1949: 11). The monogamous pair bond thus became the universal evolutionary blueprint, uniquely designed and tailored for the regulation of human sexuality, the division of labor, and the nurturance and enculturation of offspring. Similar sentiments were expressed by leaders of the structuralist-functionalist movement and helped to establish a prominent and enduring evolutionary role for the nuclear family in a generation of textbooks.[3]

As discussed in earlier chapters of this book, theories that link hominin evolution to the abandonment of polygynandrous mating and its replacement by pair bonding typically do so at the expense of ancillary baggage. Namely, corollary assumptions are made regarding disparate sexual libidos, parental investments, and intergenomic conflict. The nuclear family model provides a perfect match with traditional Western European notions about the *natural* juxtaposition of the sexes, including patterns of social dominance, sexual appetites and jealousies, cuckoldry, family provisioning, conflicting reproductive agendas, and mutual exploitation. Notably, the nuclear family is the core structure on which both the hunting hypothesis and selfish-gene theories of hominin evolution rely. In essence, the argument is that without male philopatry, mate guarding, paternity assurance, and male provisioning, the genus *Homo* would have never emerged.

Ryan and Jetha (2010), in their insightful book *Sex at Dawn*, refer to the tendency to project modern cultural stereotypes onto ancient mating behavior as "Flintstonization." Theories incorporating these notions, they note, have become an essential part of the "standard narrative," which they summarize this way:

> The standard narrative of the origins and nature of human sexuality claims to explain the development of a deceitful, reluctant sort of sexual monogamy. According to this oft-told tale, heterosexual men and women are pawns in a proxy war directed by our opposed genetic agendas. The whole catastrophe, we're told, results from the basic biological designs of males and females. Men strain to spread their cheap and plentiful seed far and wide (while still trying to control one or a few females in order to increase their paternity certainty). Meanwhile, women are guarding their limited supply of metabolically expensive eggs from unworthy suitors. But once they've roped in a provider-husband, they're quick to hike up their skirts (when ovulating) for quick-and-dirty clandestine mating opportunities with square-jawed men of obvious genetic superiority. It's not a pretty picture. (2010: 25)

Ryan and Jetha go on to question the universality of monogamous mating, pair bonding, and the nuclear family unit by documenting examples in the ethnographic record where sexuality, parenting, and provisioning are accomplished communally. They point to human societies in which multiple sexual liaisons for both males and females are not only permitted, but openly sanctioned or encouraged (in fact, insemination of females by multiple males may even be regarded as key to fertility and to successful pregnancies). In such cultures, paternity certainty and the mandate for paternal provisioning are essentially nonissues, since both the mother and her offspring typically fall under the protection and socioeconomic largess of their uterine kinsmen. In cases where the multiple sexual partners of a woman are publicly recognized as potential or equal progenitors, a custom referred to as "partible paternity," the mother and offspring may benefit both socially and materially from a community network of "many fathers" (Hrdy 2000).

Lest the proponents of universal monogamy (in its serial or more resilient forms) dismiss polygynandrous societies as exceptional or aberrant outliers, Ryan and Jetha probe more deeply into the comparative anatomies of ancient and modern primates for evidence of their past mating practices. In brief, hominin traits thought to be indicative of ancient polygynandrous mating patterns include nominal body dimorphism, the comparative size and performance of male genitalia, female ovulatory crypsis and sexual receptivity, female orgasmic capacity, and internal sperm competition.[4]

The larger question is how and why polygynandrous mating systems may have provided a selective advantage to evolving hominins. Here we are reminded that the primary reproductive and provisioning burdens among nonhuman primates fall to females. They bear the responsibility for developing social strategies that provide access to both food resources and mating partners while maintaining a margin of safety for themselves and their young. As we have seen, female primates distribute themselves spatially in relation to resource availability and often rely on intrasexual networks for structured access to both preferred foods and mates. Also, as noted by Hrdy (2000), females utilize their sexuality to create beneficial relationships with males, both for arranging favorable copulatory opportunities and enhancing the future protection of themselves and their offspring from hostile acts by male conspecifics. Hrdy proposes that polygynandrous mating provides females with a means of cementing friendly relationships with multiple, potentially aggressive males while simultaneously confusing biological paternity, a strategy proposed to reduce the vulnerability of her offspring to infanticide.[5]

As discussed in earlier chapters, ancient hominin females faced increased energy demands associated with family provisioning in terrestrial habitats and the lengthening periods of offspring dependency that accompanied encephalization. These challenges were met by extending food sharing and nurturant functions intrinsic to the matricentric unit across broader cooperative social networks. Alloparenting emerged as a strategy for reallocating the energy budget of hominin communities to benefit mothers and offspring. Hrdy argues that cooperative breeding represents the only feasible means by which mothers in both ancient and modern hunter-gatherer societies could successfully raise successive offspring to the age of economic independence. In short, "it takes a village." Hominins evolved not as pair-bonded, but as *group-bonded* primates. The standard-narrative portrait of ancient human communities as a collection of reproductive dyads—of male providers and stay-at-home moms—is no more plausible for Pleistocene populations than it is for contemporary hunter-gatherers. Hrdy comments:

> The point is not that males don't matter, but rather that there was never much evidence or a very strong theoretical basis for assuming that mothers in the Pleistocene could count on fathers to give a high priority to provisioning children they already have rather than seeking additional mates any more than mothers today can count on them to do so. (2000: 85)

It has been argued here that polygynandrous mating is an ancient hominin pattern that served to advance the reproductive agendas of both sexes. For females, liaisons with multiple males helped to establish amicable intersexual alliances and the partible paternity of offspring, both of which served to enhance the fecundity and well-being of matricentric and allopatric units. For males, such intersexual alliances provided enhanced reproductive access to females, which became increasingly based on cooperative relationships rather than aggressive behaviors.

If polygynandrous mating patterns are endemic to the genus *Homo*, how is the prevalence of pair-bonded social institutions and male-centered social groups among modern humans to be explained? Hrdy notes the importance of considering the extent to which these patterns are a comparatively recent artifact of massive population shifts in the post-Neolithic era, which saw the introduction of agriculture, pastoralism, higher population densities, and the increasing importance of portable, heritable, and defensible property:

> The question then becomes at what point in human evolution and history did patrilineal interests start to prevail? Are the consequences now inscribed in the genome of our (by bonobo standards) relatively chaste and extremely modest species? Or did evolution produce females more sensitive in this respect so that they could adapt quickly to local circumstances and customs that have long varied, and still do vary? This is one of the areas of mate preferences that we know least about, all too often overlooked in our eagerness to document essential male-female differences or to demonstrate just how "natural" patriarchal arrangements are. Yet just as social scientists cannot hope to understand human affairs without taking into account evolution, I am convinced that evolutionists cannot do so without taking into account history. (2000: 92)

Ryan and Jetha argue further that polygynandrous mating patterns still predominate among modern humans. Statistically speaking, this is true regardless of the prevailing type of formally sanctioned marriage practices or the degree to which extramural sexual liaisons are culturally discouraged or prohibited for one or both sexes. Humans, in their view, are genetically predisposed to be sexy primates with an eye for novelty and variety. Where cultural taboos intervene in the expression of such proclivities, they require extraordinary or sometimes draconian efforts to sustain, and still enjoy only limited success. In short, the traditional nuclear family as the bastion of human sexual monogamy is a cultural myth.

Behavioral Plasticity

Behavioral plasticity refers to an organism's ability to express more than one phenotype (observable behavior) from a single genotype in response to external factors.[6] Neurological specializations accompanying the evolution of the hominin brain emancipated our genus from reliance on predominantly hardwired responses to subsistence and reproductive challenges, providing a broader repertoire of flexible behaviors with which to adapt to changing environmental conditions. Plasticity is the product of a complex gene-environment interaction whereby the phenotypic expression of genes may be affected by social learning and modified at the molecular level without changing the underlying DNA structure (Mery and Burns 2010; Snell-Rood et al. 2010; Ledon-Rettig, Richards, and Martin 2012; Snell-Rood 2013). The process by which such behavioral variation in gene expression occurs is referred to as *epigenesis*, literally "extra growth."

Epigenetic traits have affected the trajectory of human evolution in two important ways. First, the ability of hominins to express different phenotypes in different environments and to vary their behavioral responses over the course of a lifetime is critical to optimal fitness. In other words, nongenetic phenotypic heterogeneity may evolve by natural selection if a distribution of phenotypes confers greater fitness than a single type. Phenotypic plasticity facilitated the expansion of populations into novel habitats, reduced their vulnerability to stochastic events, and provided a mechanism for the modification of reproductive strategies. Second, stable epigenetic traits ("epialleles") are significant in that they may lead to heritable transgenerational changes in phenotype.[7] Dickens and Rahman (2012) argue that epigenetic processes, or what are sometimes referred to as "soft inheritance systems," along with individual and cultural learning, are part of a "nested hierarchy of adaptations" on which natural selection acts to calibrate organisms with environmental stochasticity:

> What this imagined scenario brings is the view that increased ecological complexity, or large ecological band width, brings with it an increased need for highly integrated calibrating mechanisms. These mechanisms are dealing with various sources of information and fine-tuning the organization of the phenotypic response to factors that are at best described as exogenous to the genes that build them. This is plasticity, this is development and this does lead to changes in the frequencies of available phenotypes in the population. Genetically encoded capacities for generating epigenetic variation may drive part of this phenotypic plasticity. (Dickens and Rahman 2012: 7)

Epigenetic systems, because of their greater flexibility and potential for mutation, may represent a fast-track alternative for the inheritance of phenotypic variation, as opposed to the slower, more stable modifications in DNA sequences (Bossdorf, Richards, and Pigliucci 2008; Richards, Bossdorf, and Pigliucci 2010). Such systems could potentially serve as a vehicle for multilevel selection under conditions of rapid environmental change, enhancing not only individual fitness but conferring adaptive advantages at the group level. As noted by Ledon-Rettig and colleagues:

> If heritable epigenetic variation plays a role in adaptation, then local differences in habitat characteristics may select for different epialleles in different populations. As with genetic polymorphisms, this selection will result in population-level associations between epialleles, behaviors, and ecological habitat characteristics. (Ledon-Rettig, Richards, and Martin 2012: 313)

An analysis of behavioral plasticity (flexibility) among contemporary primates was conducted by Clara Jones (2005), who examined the relationship of phenotypic variance with inclusive fitness, ecology, and environmental change. An advocate of intergenomic conflict models, Jones sees male and female primates as pursuing distinct and opposing reproductive strategies. Females, whose fitness is based on food and the nurturance of offspring, must compete intrasexually for access to resources while simultaneously avoiding parasitism by males. Their behavioral strategies thus reflect the relative costs and benefits of choices that maximize energy. Males, who are largely emancipated from energy investments in reproduction, instead compete intrasexually for access to and exploitation of fertile females. Since male fitness rests on the frequency of successful copulations in competition with conspecifics, their strategies reflect the relative costs and benefits of coercive or dominance behaviors that minimize time. Reliance on cognitive mechanisms such as learning and Machiavellian intelligence provide primates with critical information to guide their decision making and the exercise of flexible phenotypic behaviors.

Jones notes that behavioral plasticity is most prominent among primates living in heterogenous environmental regimes, conditions that also favor assemblages of large multimale-multifemale groups and polygynandrous mating systems. It is within such sociosexual organizations that the competing reproductive interests of the sexes are played out and that novel behaviors emerge for their resolution. In large multimale-multifemale polygynandrous groups, males have expanded opportunities for sexual access and achieve greater fitness by exhibiting aggressive restraint and cooperative behaviors toward females. In other words, the costs of intrasexual competition and intersexual coercion outweigh the benefits. Similarly, females pursue altruistic or cooperative behaviors such as social foraging, allocare, and sexual selection to the extent that such behaviors maximize their energy budgets. Jones goes on to suggest that, due to the nexus of foraging and sociality with positive energetic effects, females may be the principal drivers of social evolution in primates.

In summary, primate behavioral plasticity is a reflection of complex interactions between genetics, epigenetics, and ecology. Observations of contemporary primates illustrate how the ability to respond to novel conditions with flexible behaviors enhances the fitness of both sexes, and hence why it may have been favored by natural selection. The behavior of ancient hominins probably diverged from that of contemporary nonhuman primates in two important ways: brain evolution and eusociality. As discussed below, the linear up-scaling of the primate brain is believed to have been accompanied by a morphological and functional reorganization of its component structures, the net effect of which was increasing cortical modulation and control over more primitive limbic systems. These changes would suggest a higher frequency of behavioral plasticity among ancient hominins over time based on the integration of perceptual, learning, and cognitive functions. The second point of divergence lies in the evolution of eusociality, in which the reproductive interests of individuals become increasingly aligned with, and in some cases subordinated to, the interests of the group. For hominin populations, inclusive fitness is more than a zero-sum game where the success of one sex occurs at the expense of the other. Behavioral plasticity provided ancient hominins the wherewithal to advance individual reproductive success through the establishment of cooperative relationships and corporate activity among group members. Human evolution thus involved the complex interplay of natural selection at both the individual and group levels.

Encephalization and Intelligence

The ability of ancient hominins to conceptualize a three-dimensional roadmap of their physical and social worlds and to organize themselves into collaborative groups for subsistence and reproduction presumes a capacity for abstract thought and some form of effective communication. Scholars sitting atop the phylogenetic tree, however, have historically resisted this notion, reserving such traits and membership in the exclusive club called humanity for Late Pleistocene peoples. Because phylogenetic models have typically calibrated the development of cognitive and language abilities with increases in absolute brain size, it has been argued that humans essentially did not really "arrive" until the last phases of the Ice Age.

Evidence of cognitive, altruistic, and moral behaviors exhibited by our distant primate cousins, both living and extinct, is frequently discounted, a tendency referred to by de Waal (2016: 22) as "anthropodenial." Claims for the exceptionalism of modern humans, however, are being increasingly challenged by new paleontological and archaeological data, by social intelligence theory, and by groundbreaking neurological research on primate brain evolution. These perspectives will be briefly reviewed.

Pleistocene Material Remains

As noted earlier in this chapter, it is common for evolutionary reconstructions to begin with the Terminal Pleistocene and dial the clock backwards. If, it's reasoned, anatomically modern humans (AMHs) spread at the expense of other hominin lineages and laid the foundation for more complex cultures, it must mean that they were fundamentally different primates, imbued with superior physical traits, intelligence, and foresight. *Homo sapiens sapiens* has traditionally been designated as the progenitor of "modern" cultural behavior, as evidenced by the species' advanced toolkits and elements of symbolic culture, such as language, music, art, and religion. These traits have traditionally been characterized as a sudden evolutionary development—the so-called "Great Leap Forward" or "Upper Paleolithic Revolution" that coincided with the late geographical expansion of AMHs from Africa into Eurasia around 60 ka. The theory that modern humans dispersed from Africa as a single wave in the Late Pleistocene, however, has been called into question by recent paleontological and genetic evidence. The more contemporary view is that AMHs evolved on that continent millennia earlier and radiated into Eurasia in a series of dispersals, interbreeding with Neanderthals, Denisovans, and perhaps earlier hominin lineages as they went (Bae, Douka, and Petraglia 2017).

Assigning credit for cultural complexity to the extraordinary abilities of AMHs raises what Colin Renfrew (2007) referred to as the "sapient paradox." Namely, how to explain the significant time gap of as much as a quarter million years between the appearance of early AMH fossil remains in Africa and the subsequent emergence of advanced Upper Paleolithic and Neolithic cultures in Europe. If there was a sudden change in the cultural trajectory of our species, what was the catalyst?

One proposed solution to the paradox is that the efflorescence of technical and cultural complexity witnessed at the end of the Ice Age is attributable to a late mutation in the sapient brain that occurred between one hundred thousand and fifty thousand years ago (Mithen

1996; Coolidge and Wynn 2009; Klein 2001; Barnard 2011). This theory has also been adopted in recent popular works such as Harari's *Sapiens,* which credits a "tree of knowledge" mutation to the species' "cognitive revolution" (2011: 23–24).[8] No corroborating DNA evidence, however, has been advanced for such a mutation.

Other solutions to the paradox link Late Pleistocene complexity to changing ecological rather than genetic factors. For example, Powell, Shennan, and Thomas (2009) propose that both technological innovation and social complexity in AMH populations were triggered by their migrations into new environments and the accompanying effects of increased subpopulation densities on the accumulation and transmission of cultural knowledge. Similarly, Wengrow and Graeber (2015) suggest that seasonal aggregations of AMH populations living on the glacial fringe of Upper Paleolithic Europe created enhanced opportunities for the elaboration of sociopolitical structures and symbolic expression. In other words, favorable demographics rather than increased cognitive capacity account for the timing and geographical appearance of increased technological and symbolic complexity.

Other theorists have refuted the sapient paradox notion entirely by arguing that the emergence of so-called modern behaviors was not a sudden evolutionary event, but the culmination of a gradual process extending over hundreds of thousands of years. McBrearty and Brooks (2000), in their comprehensive analysis of African fossil and archaeological records, argue that the elements of modernity arose at different times and in different regions on the African continent during the Middle Stone Age, millennia before their export to Europe. The authors counter past theories that limit the emergence of cultural complexity to Upper Paleolithic Europe by citing evidence of germinal modern behaviors at selected African Middle Stone Age sites. They propose that a plausible ancestor of AMHs (*Homo helmei*) could extend evidence for the continuous evolution of modern human biocultural traits in Africa to the beginning of this period, over a quarter million years ago. The recent discovery of early AMH paleoarchaeological remains in Morocco dated 300 ka lends some credence to this theory (Stringer and Galway-Witham 2017). McBrearty and Brooks conclude that there was no sudden change or difference in the cognitive abilities of African Middle Stone Age and Late Stone Age hominins.

Notably, however, no such intellectual continuity has been proposed for the contemporaneous evolution of hominins living in contiguous areas just across the Strait of Gibraltar and on the Arabian Peninsula during this same time frame, i.e., between 350 ka and the radiation of Neanderthals throughout Eurasia. The presumption seems to be that no gene flow occurred across the Mediterranean over the ensuing millennia or that, as Harari (2011: 24) proposes, Neanderthals were simply excluded from the gene pool that, by chance, received the "tree of knowledge" mutation. Theories on the nature and extent of AMH exceptionalism, while subject to periodic revision, have tended to bias the way in which other hominin lineages are perceived.

For example, Neanderthals have historically been stereotyped as brutish, dull-witted, and otherwise lacking the cognitive wherewithal to adapt to changing conditions in the Late Pleistocene. Early theories perceived them as so unlike us that, despite their large brains, they were cast as a distinct species without complex language or the ability for abstract thought. While some recent characterizations see Neanderthals as capable of rudimentary speech and simple kinship systems, they are still frequently portrayed as falling far short of the cognitive abilities and achievements reserved for AMHs as the standard bearers of a latecomer "symbolic revolution" (Barnard 2011).

Recent excavations of early Western European sites, however, have added new wrinkles to the long-standing debate on the capacity of Neanderthals for symbolic behavior. For example, cave art at three sites in Spain has been dated by uranium-thorium techniques at 68.4 ka, or approximately 20 millennia prior to the purported entry of AMHs into the Iberian Peninsula, one of the last European strongholds of Neanderthal populations (Appenzeller 2018; Hoffmann et al. 2018). Other investigators have pointed to much earlier evidence of symbolic behaviors at Bruniquel Cave in southwest France (Jaubert et al. 2016). Excavations at this site, dated at 176.5 ka, yielded circular structures constructed of broken stalagmites located deep within the cave and associated with fire and burnt bone. These and other recent discoveries suggest that so-called "modern" cultural behavior has a much greater antiquity than previously supposed and that its emergence was a parallel development among evolving hominins on both the African and European continents.

Portraits of Neanderthals in traditional theory as our less-gifted distant cousins also cast them as easy targets for displacement by more nimble and intelligent hominins. Indeed, some of the more colorful accounts of AMH expansions, such as the spread of Gravettians into Europe, might well be set against a musical backdrop of Wagner's "Ride of the Valkyries." More recent models are somewhat kinder in pointing to the underpinnings of perceived Neanderthal competitive disadvantages. Finlayson (2009), for example, opines that Neanderthal extinction may be attributable to their body morphology—to their squat build and short legs that left them ill-equipped for hunting in the increasingly treeless landscape of the Late Pleistocene. This morphology is contrasted with the linear, long-legged AMH body type that evolved in arid climates, an allegedly superior hunting physique that Finlayson notes might have also been enjoyed by Neanderthals had they evolved under similar ecological conditions. But were Neanderthals simply bipedally challenged?

Somewhere in such evolutionary scenarios, lessons so eloquently enumerated by Jared Diamond (1997) about the varying complexity and standing of human populations through time and space have been ignored. Similar to the stark contrasts that could be drawn between modern industrialized societies and surviving hunter-gatherers, such differences in sociocultural complexity are not indicative of relative intelligence, but rather are the culmination of both happenstance and historical externalities.

At the end of the Pleistocene, Neanderthals were practicing cultural traditions in their ancient homelands that had been successful for millennia. The old ways of doing things, however, were being increasingly challenged by the effects of climate change. Landscapes and faunal communities were being rapidly transformed. But just as necessity is the mother of invention, so too is certainty the mother of investment. Chris Low (2017), in his critique of the cognitive revolution concept, notes that the conservatism of human groups in the face of change should not be dismissed as a lack of ingenuity:

> Our very distant ancestors performed with an aesthetic intelligence rooted in feelings for what is right. To say that a static archaeological record, like that found in phases of the pre "revolution" Stone Age, equates to creative and intellectual doldrums is to mistake progress for evolution, the latter being nothing more than change into which we read direction. Hunter-gatherers know you do not waste time and amplify risk by changing something unless you really have to. (2017: 244)

Evidence suggests that Neanderthals and AMHs lived side by side in some regions for about ten thousand years, at times maintaining their separateness and at other times sharing both their cultural traditions and their genes. In contemplating the fate of Neanderthals, therefore, it seems unnecessary to equate their eventual disappearance as a distinct lineage with extinction, nor to equate this outcome to diminished cognitive prowess. It is just as reasonable to conclude that they were culturally and genetically absorbed by incoming AMH peoples, whose productive success was spurred by an unprecedented convergence of innovations in material technology, favorable turns in climate, increased fecundity, and demographic changes that fostered greater labor efficiencies and sociopolitical complexity.[9]

Thanks to the recent sequencing of the Neanderthal and AMH genomes, the darker portraits of our big-brained cousins have been subject to challenge, if not set aside. The presence of the FOXP2 "language gene" has now been established in the Neanderthal genome (Krause et al. 2007). Middle Paleolithic archaeological sites provide other indirect indications of cognitive abilities and symbolic behaviors, as evidenced by the presence of burials; art objects such as the 230 ka carved Venus figure from Berekhat Ram, Israel; the building of shelters; and the frequent association of Neanderthals with coordinated game drives and large kill sites (Goren-Inbar 1986; Schepartz 1993).

The extraordinary finds recently unearthed at Schöningen, Germany, suggest that complex social and economic behaviors extend further back in time to Neanderthal ancestral lineages (Conard et al. 2015). This 300 ka *H. heidelbergensis* butchery site contains the skeletons of twenty to twenty-five horses, along with wooden spears and lithic assemblages. These remains have been interpreted as evidence of cooperative hunting that would have required considerable advanced planning and communication. Additional isotopic analyses of the skeletal remains at the site concluded that the slain animals came from different populations that died at different times of the year. This finding supports the hypothesis that this location hosted multiple kill events and may have served as a hot spot at which hominin groups periodically aggregated to not only conduct communal hunts, but also perhaps to gather and share a variety of other resources available in this lakeshore habitat, such as fish and underground storage organs.

An even earlier Middle Pleistocene *H. heidelbergensis* butchery site, located at Boxgrove in the English county of West Sussex, is dated at 500ka. This site is situated at the foot of a buried chalk cliff that overlooked a flat beach and waterhole that extended about a half mile south to the sea. Boxgrove is notable for the discovery of a *H. heidelbergensis* tibia and teeth, along with a large inventory of worked flint tools and a wealth of well-preserved bones of mammalian fauna (horse, bison, large deer, bear, and rhinoceros) and smaller species (birds, frogs, and voles).

The coordinated socioeconomic activities implied by remains at the Schöningen and Boxgrove sites suggest the presence of social networks and sophisticated forms of communication. While the presence of language is only inferred by such behaviors, some investigators have also pointed to evolutionary changes in genes, such as EYA1, associated with anatomical structures of the outer and middle ears in hominins of this period. For example, *H. heidelbergensis* skulls from the Sierra de Atapuerca site in Spain, dated 350 ka, were found to reflect evolutionary

Figure 5.1. Middle Pleistocene *Homo heidelbergensis* butchery site at Boxgrove, West Sussex, England. With permission of Getty Images.

changes in these structures compatible with the human sound transmission pattern and hearing acuity necessary for understanding the spoken word (Martinez et al. 2004).

The penchant for "dumbing down" early human populations tends to accelerate as theorists proceed down the family tree. This perceptual lens is particularly problematic for more ancient lineages, such as *H. erectus,* that have volumetrically smaller brains and appear to have relied heavily on scavenging and foraging rather than organized hunting. Some observers have taken this to mean that such hominins followed a kind of rote or serendipitous approach to food getting and environmental adaptation that required neither collaboration, labor division, nor foresight. For example, Boaz (1997: 177) opines that the *H. erectus* brain size suggests mental abilities equivalent to that of a modern five-year old. Such conclusions conjure images of ancient society operating on a level akin to *Lord of the Flies*.

In fact, *H. erectus* has been characterized not only as a dullard, but as a "bullnecked and bullet-headed species" and "blood-thirsty killer" by Boaz and Ciochon (2004: 108, 171), who interpret the cranial pachyostosis (thick-boned skulls) of some *H. erectus* fossil specimens as a defensive adaptation to chronic head bashing. In their view, the 1.5-million-year career of *H. erectus* was marked by interpersonal violence so constant and so prolonged that it led to the evolution of a "hard hat" cranium. Such genocidal tendencies are attributed to competition for mates, and its morphological consequences to sexual selection. Boaz and Ciochon also opine that intraspecies violence was a constant occurrence between small, insular *H. erectus* bands due to climate-induced impacts on territorial ranges and resources. In an earlier publication, Boaz concluded: ". . . this head-bashing, cannibalistic, cave-dwelling, smoky-smelling, and slavishly imitative hominin was a far cry from the concept of 'human'" (1997: 189).

Such intuitive caricatures overlook the fact that *H. erectus* lineages expanded across Africa and Eurasia, successfully adapting to a wide range of habitats and surviving climate change challenges for over 1.5 million years. This is not only an impressive undertaking, but arguably a demonstration of foresight, ingenuity, collaboration, and perseverance. It is a journey that has left only a sparse material record, but one whose success would likely have been compromised by endemic intraspecific violence. As proposed in chapter 3, competition for sexual access was a drama that was played out and probably resolved early in hominin evolutionary history in the context of cooperative, multimale-multifemale breeding communities. Similarly, as noted by Finlayson (2004: 153), the conditions for resource competition arise when competing populations are at carrying capacity and their respective need and use of such resources impacts the growth potential of the other. Given the probable densities of *H. erectus* populations, such conditions were unlikely to have arisen frequently or to have persisted for millennia. Intraspecific warfare, in fact, is generally thought to have been rare in the Paleolithic. Recent analyses of Asian and African *H. erectus* skulls by Copes and Kimbel (2016) conclude that their cranial vault thickness is not unique among primates, thereby casting further doubt on the "head bashing" hypothesis.

Insights into the cognitive and symbolic life of *H. erectus* are challenging to glean from the limited material remains at hand, which consist largely of simple stone tools. Several investigators have studied Mode 1 (Oldowan) and Mode 2 (Acheulean) tool-making techniques and their potential implications for intelligence and language. Toth and Schick (1993) concluded that the simple Mode 1 tools were produced more on an ad hoc basis than on a preconceived design. In contrast, they saw Mode 2 tools as the result of a more sophisticated process, wherein the tool maker proceeded with a mental image of the final product and the necessary techniques to produce them consistently. Wynn (1993) reached similar conclusions, arguing that the production by *H. erectus* of the prototypical Acheulean biface handaxe required a higher level of cognitive function than was indicated for the makers of Mode 1 tools, such as *H. habilis*. He proposes that this greater cognitive requirement and the uniformity of Acheulean tool types across much of Africa and Eurasia is indicative of the social transmission of Mode 2 stone-knapping techniques.

More recently, Morgan et al. (2015) investigated Oldowan and Acheulean tool making within the framework of gene-culture evolutionary theory. They assumed that both technologies required skills that were learned and socially transmitted and that the fitness benefits of these tools in turn created selective pressures for teaching and language. They conducted a large-scale experiment to investigate the efficacy of various social learning mechanisms to transmit Oldowan knapping techniques. Their findings were that simple imitation or emulation had little effect on performance, whereas teaching mechanisms and particularly verbal instruction rapidly increased the quality of flaking. They concluded that selection would have favored an early origin for language and that this capacity for symbolic communication evolved slowly over time. The stasis observed in Oldowan technology for about seven hundred thousand years, they propose, is explained by "low-fidelity" forms of social transmission, whereas the greater complexity of Acheulean technology after 1.7 ma signals the emergence of "high-fidelity" communication involving symboling and rudimentary forms of language.

The link between Acheulean technology and symbolic communication has also been made through the interpretation of archaeological remains. For example, Rabinovich, Gaudzinski-Windhauser, and Goren-Inbar (2008), in their excavations of Acheulean layers at the early Middle Pleistocene site of Gesher Benot Ya'aqov, Israel, found an early example of fallow deer butchering so systematic and so anatomically precise that a shared strategy and knowledge base was assumed: "We interpret the Gesher Benot Ya'aqov data as indicating that the Acheulian hunters at the site (1) were proficient communicators and learners and (2) possessed anatomical knowledge, considerable manual skill, impressive technological abilities, and foresight" (2008: 134). The investigators noted that this butchery site resembled those of much later Upper Paleolithic examples.

While the degree of sophistication exhibited in stone knapping technology has been proposed as a marker of advanced cognitive abilities, the reverse assumption is not necessarily true. For instance, there are some areas of the Old World, such as Southeast Asia, where Oldowan traditions persisted and the prototypical Acheulean handaxe never appears in the archaeological record. Similarly, at the *H. heidelbergensis* horse butchery site in Schöningren referenced earlier (Conard et al. 2015), lithic inventories included only flake tools, while Acheulean handaxes were commonplace at the earlier Boxgrove site. A cautionary note, therefore, is that Mode 1 and Mode 2 technology types may, in some instances, simply reflect their usefulness in a given niche or to the task at hand, rather than the mental aptitudes of their makers.

The presumed relationship among knapping technologies, intelligence, and symbolic behavior is even more difficult to decipher when interpreting the significance of Oldowan tool making among the more ancient *H. habilis* and australopithecine populations. One of the vestiges of the "hunting hangover" is the lingering theory that cognitive development went hand in hand with organized hunting by males. Current evidence, however, suggests that stone tool making evolved much earlier, and that tool use was associated with a wide range of subsistence endeavors. This begs the question of whether reliance on stone tools and the social transmission of their manufacturing techniques was the primary selective factor in the evolution of intelligence and language, or whether, instead, there was something about the fitness advantages of behavioral plasticity and adapting to life in cooperative socioeconomic groups that itself favored the advancement of cognitive and communication skills.

Tanner (1981: 206–8), citing the analyses of cranial endocasts by Holloway (1972a, 1972b, 1976), suggested that reorganization of the cerebral cortex in regions designated as "association areas" appears to have already begun in australopithecines. Endocasts, unfortunately, cannot provide a blueprint of brain circuitry. Neither, perhaps, can brain size or volumetric estimates of the neocortex be relied on as accurate predictors of complex function. Does quantity always calibrate with quality? The recent *Homo naledi* finds in South Africa beg the question of how fifteen fossilized individuals ended up in a nearly inaccessible cave chamber (Berger et al. 2015). Could ancient hominins with human-like bodies but the cranial capacity of an australopithecine be capable of the symbolic behavior typically associated with special treatment and deposition of the dead? The discovery of these skeletal remains and their relatively recent provenance serves as a reminder of how much is yet to be learned about the sheer variety and evolutionary journey of past hominin lineages, how they lived, and how their minds worked.

While future fossil and material discoveries will shed further light on the richness and diversity of ancient life, the precise origins of language and complex reasoning will probably remain a mystery. What can be said with some confidence, however, is that members of the *Homo* family tree had a running start. Hominins entered the Late Pliocene equipped with cognitive abilities and other prosocial traits, along with the skills to make simple flake tools. The essential ingredient that separated our ancestors from other primate lineages, however, was probably not the invention and transmission of tool-making techniques, but the invention and transmission of *social* technologies—the ability to organize themselves into sustainable, cooperative breeding communities that enhanced the reproductive success of individual members and the groups of which they were a part. It was within the boundaries of such communities that autocatalytic and co-evolutionary processes occurred and that signature cognitive and communication skills were honed.

Neocortex Size and the Social Brain

The evolutionary pathway of hominins has been marked by a gradual increase in absolute brain size. Over a century ago, pioneering studies by Brodman (1912) concluded that one region of the human brain, the prefrontal or neocortex, was disproportionately larger than would be expected for a primate of similar body weight. The association of this brain region with cognitive function generated scientific interest in the potential significance of neocortex size for the evolution of intelligence. Subsequent theories have linked neocortical expansion through time to the size and complexity of hominin social groups, as well as the capacity for language and culture.

A central tenet of such theories is that hominins evolved big brains or big neocortices due to selective factors arising out of the cognitive demands of group living. Behavioral plasticity and social networking skills went hand in hand with inclusive fitness. During his initial observations of chimpanzees in 1975, primatologist Frans de Waal was struck by the importance of hierarchies and competitiveness in their social interactions, reminiscent of the political intrigue in Machiavelli's *The Prince*. When, in a subsequent publication, he described chimp behavior as "Machiavellian," the branding of a hypothesis of the same name was born (de Waal 2007, orig. 1982; 2016: 167–77).[10] As generally applied, the Machiavellian Hypothesis (also referred to as "political intelligence" or "social intelligence") holds that the secret to an individual's reproductive success rests on their ability to successfully engage in primate politics—to apply their innate abilities for mind reading, deception, and manipulation to create beneficial relationships and outcomes vis-à-vis other group members.

British anthropologist Robin Dunbar (1988, 1992, 1995, 2003) married the concept of Machiavellian intelligence with neocortex size in his Social Brain Theory, casting both in an evolutionary context. Dunbar's theory is that the physiology of evolving brains is a double-edged sword. That is, the size of the neocortex gradually increased over time, but along the way also served as a limiting factor on the size and complexity of human groups by imposing a cognitive ceiling on the number of social relationships that individuals could effectively maintain. Dunbar based this hypothesis on comparative studies conducted with anthropoid apes, in which neocortex volumetric calculations were correlated with the size of contemporary social groups.

This same group size hypothesis was subsequently extended to extinct hominins (Aiello and Dunbar 1993; Dunbar 1993). Dunbar predicted the average social group size of ancient taxa by estimating the neocortex volume of various fossil specimens through cranial endocasts and then applying the neocortex-group size ratios previously calculated for living hominoids. By these measures, the average size of ancient social groups was proposed to be 65–70 for australopithecines, 75–80 for *Homo habilis*, 110 for *Homo erectus*, 120–130 for Archaic *Homo sapiens*, 140 for Neanderthals, and about 150 for AMHs. Assuming a nexus between neocortex size and social networking tools, Dunbar concluded that language was entirely absent in *H. erectus* and earlier hominins and appeared in only rudimentary form among Neanderthals. According to social brain theory, grammatical speech and the advanced level of cognition required for "high culture" and religion was limited to AMHs and the so-called Upper Paleolithic Revolution. Similar conclusions have been reached by Gamble (2008) and Barnard (2011), based largely on Dunbar's work. Further consideration of the demographic implications of potential proto-language or grammatical speech among early hominins appears in the discussion of mosaic brain evolution below.

Other investigators have attached special significance to neocortex size for the evolution of cognitive and social abilities. For example, Reader and Laland (2002) compared neocortex size among 115 primate species with the frequency of documented examples of social learning, innovation, and tool use, or what they referred to as "ecological intelligence." Interspecific performance data were found to correlate positively with the executive brain ratios of their subjects (as measured by neocortex volume over brain stem volume), but not with Dunbar's variable of group size. The authors offered an alternative to the social brain hypothesis, suggesting that brain evolution was shaped by multiple selective factors that favored behavioral plasticity and the ability to engage in successful problem solving in both the ecological and social domains.[11]

As with all statistical studies, correlative relationships between neocortex size, group size, and other variables are indicative of association, but not necessarily of cause and effect. The interpretation of comparative neocortex study findings has also been complicated by the inconsistent methods by which this brain region has been measured (i.e., as a volumetric ratio over body size, body weight, brain stem volume, or total brain size), and by the relative accuracy of neocortex measurements that may be derived from fossil cranial endocasts.

The conclusion that symbolic communication was limited to comparatively recent big-brained humans has been challenged by evolutionary psychologist Steven Pinker (1994: 340–81), who proposes that the capacity for language is an instinctive trait that could have appeared very early in hominin evolution. Pinker discredits the alleged nexus between language origins and expansion of the neocortex, arguing instead for the importance of changes in brain microcircuitry rather than gross brain size. Recent neurological studies are lending credence to Pinker's theory. A significant challenge to the social brain hypothesis is being posed by new information on how the primate brain has scaled-up through time. The long-standing, underlying assumption that human evolution was the result of disproportionate neocortex growth is being called into question.

Mosaic Brain Evolution

Recent neurological studies have shifted the focus of investigations on hominin encephalization from the neocortex per se to an analysis of the entire brain—to the antecedents and consequences of brain *reorganization* through time. The primate brain contains a number of distinct component systems that are the product of mosaic evolution. That is, natural selection has acted separately on the various structures and functions of the brain during the course of evolution, resulting in disparate patterns of neuronal allocation and connectivity among and between its component parts. Structures that are anatomically and functionally linked co-evolved independently, creating complex relationships among individual brain components (Barton and Harvey 2000; Smaers and Soligo 2013).

One of the more significant new findings of neurological investigations is that primate brain regions scale in a linear fashion with whole brain size and body size (de Winter and Oxnard 2000; Passingham 2002; Semendeferi, et al. 2002). Semendeferi et al. utilized magnetic imaging of the brains of living human and nonhuman primate subjects to obtain frontal cortex volumetric measurements. They found that, notwithstanding differences between apes and humans in the absolute size of the frontal cortex, the values for humans were on a par with great apes on a proportional basis (i.e., when frontal cortex volume is calculated as a percentage of the cortex volume of the cerebral hemispheres). In other words, the human frontal cortex size is *not* disproportionate to what would be expected for a primate brain of human size. The authors acknowledge that their research contradicts the original conclusions reached by Brodman on frontal cortex size. They attribute differences in their findings to the larger sample size, improved methodology (MRI cortex volumetric calculations versus surface estimates), and utilization of live subjects rather than postmortem specimens. They conclude:

> The frontal cortices could support the outstanding cognitive capabilities of humans without undergoing a disproportionate overall increase in size. This region may have undergone a reorganization that includes enlargement of selected, but not all, cortical areas to the detriment of others. The same neural circuits might be more richly interconnected within the frontal sectors themselves and between those sectors and other brain regions. Also, subsectors of the frontal lobe might have undergone a modification of local circuitry, expressed, for instance, in the form of distinct cytoarchitectonic patterns. (Semendeferi et al. 2002: 275)

A second major contribution of neurological studies to our understanding of hominin brain evolution is the central role played by changes in neuronal density and the interconnectivity of brain structures. Seminal studies conducted by Herculano-Houzel (2009) have drawn attention to the functional and co-evolutionary relationship between the cerebral cortex and cerebellum, both of which have higher neuronal densities than the rest of the brain. Her findings suggest that information processing and intelligence in all species is linked not to brain size per se, but to the number and type of cortical neurons, circuitry, and conduction velocity or speed. Primate brains are distinguished by the economical manner in which they have accommodated neuronal expansion without an increase in size:

> Quantitative changes in the neuronal composition of the brain could therefore be a main driving force that, through the exponential combination of processing units, and therefore of computational abilities, lead to events that may look like "jumps" in the evolution of brains and intelligence. Such quantitative changes are likely to be warranted by increases in the absolute (rather than relative) numbers of neurons in relevant cortical areas and, coordinately, in the cerebellar circuits that interact with them. (Herculano-Houzel 2009: 8)

Herculano-Houzel, Manger, and Kaas (2014), later examined the neuronal scaling rules applicable to the evolutionary branching patterns of six mammalian clades, including primates (fifteen species, including humans). While their detailed findings go beyond the scope of this book, general conclusions on primate brain evolution provide insights into how hominid social intelligence emerged. Important in this regard is the observation that the cerebral cortex and cerebellum have higher scaling ratios of neurons over the rest of the brain, and that interconnections between such structures "add a whole new level of elaboration to the processing of information relayed from the body and back to it through associative processing, endowing animals with more refined and flexible behavioral repertoires" (2014: 15).

Associative activity involving the simultaneous activation of neurons is thought to increase the synaptic strength between them and provide the foundation for the emergence of neuronal networks. For example, Carillo-Reid et al. (2016) demonstrated in a recent chimera study that coactive groups of neurons constitute neuronal ensembles that, with recurrent activation, increase in functional connectivity and become spontaneously coactive and imprinted in the circuit. This basic theory in neuroscience, referred to as cell assembly theory or "Hebb's Rule," was originally proposed by Donald Hebb (1949) as an explanation for brain adaptations during the learning process. It also provides a theory for the emergence of so-called "mirror neurons." These neurons are activated not only when an individual is performing an action, but also when that individual observes such an action performed by others. A central concept of Hebb's Rule is that "neurons that wire together fire together." In other words, the same sensory neurons and synapses involved in performing an action will, upon repetition, become imprinted such that they will begin auto-firing when stimulated by the perception of a similar action by others. Mirror neurons are considered to be central to the experience of empathy among apes and humans, a characteristic feature of primate sociality.

Recent research involving another class of brain cells, von Economo or "spindle" neurons, has extended the application of Hebbian principles to the evolution of social intelligence (Nimchinsky et al. 1999; Allman et al. 2001; Allman, Hakeem, and Watson 2002). Spindle neurons are found in only a limited number of large-brained primates, elephants, and cetaceans and are associated with high-velocity connectivity between the anterior cingulate cortex (ACC) and insular cortex and other parts of the brain. Allman's research team found spindle neuron relays between the ACC and Brodmann's area 10 of the frontal polar cortex, a region known to regulate cognitive dissonance. They theorize that the ACC serves as a gatekeeper for emotions, rapidly relaying signals received from the amygdala (the emotion processing center) to Brodmann's area where memories of past experiences are retrieved and compared, and then provide the basis for formulation of adaptive responses to novel conditions. These neurobehavioral specializations are thought to be integral to self-control, rational decision making, and social cognition and provide a critical piece to the puzzle on how behavioral plasticity, epigenetic traits, and human sociality evolved.[12]

An additional finding of this research is that humans and great apes differ in the average quantity of spindle neurons they have, as measured by the number per section present in the ACC region of their brains: approximately eighty-nine for humans, sixty-eight for bonobos, thirty-seven for chimpanzees, and only nine for orangutans (Allman et al. 2002). Of interest are the potential implications of spindle neuron densities for behavior. Bonobos and chimpanzees have comparable brain sizes and a close phylogenetic relationship, but vastly different temperaments and social lives (Wrangham and Peterson 1996; de Waal 2013). Rilling et al. (2012) set out to examine the neural systems of both apes to discover if they held the clues to bonobo sociality, which is marked by less aggression and greater playfulness, sexuality, tolerance, and empathy. The research team conducted MRI brain scans on live subjects, as well as analyses of postmortem specimens. They summarized their results as follows:

> We find that bonobos have more gray matter in brain regions involved in perceiving distress in both oneself and others, including the right dorsal amygdala and right anterior insula. Bonobos also have a larger pathway linking the amygdala with the ventral anterior cingulate cortex, a pathway implicated in both top-down control of aggressive impulses as well as bottom up biases against harming others. We suggest that this neural system not only supports increased empathic sensitivity in bonobos, but also behaviors like sex and play that serve to dissipate tension, thereby limiting distress and anxiety to levels conducive with presocial behavior. (2012: 369)

The implications of these neurobehavioral specializations for primate brain reorganization and the evolution of intelligence are potentially far-reaching. They provide a biological basis for understanding how natural selection shaped various regions of the brain to effectively receive, interpret, and respond to changing environmental stimuli in a manner that enhanced behavioral plasticity and reproductive success. Their presence in our closest living relatives speaks not only to our common genetic heritage, but to the extent to which neurological differences may impact phenotypic behavior at the individual and group levels.[13]

What if, as de Waal (2013: 61) muses, our *Pan-Homo* progenitor was more like a bonobo than a chimpanzee? What if, over 5 million years ago, *Ardipithecus* and similar bipedal apes had already acquired neurobehavioral specializations equal to or exceeding those of bonobos that set hominins on a separate social trajectory? What if, as Pinker (1994: 340–81) proposes, the capacity for language evolved as an instinctive hominin trait that may have been present, at least in rudimentary form, among our habiline and *H. erectus* ancestors? Neurobiological research findings serve as a cautionary note that, when considering the aptitudes and lifeways of our ancient hominin forebears, we not judge a book by its cover. Cranial endocasts of our fossil ancestors may provide evidence of the linear scaling of primate brains. But they tell us nothing about the neuronal densities, spindle cells, and internal circuitry they contained, and hence the relative intelligence and communication skills of their owners.

Social Demography

Hominins define themselves by their relationship with others. They are driven to membership in social groups to not only pursue reproductive goals, but to secure the sustenance, mutual aid, camaraderie, solace, and comfort that kith and kin provide. To be banished from one's group is a dire and potentially fatal predicament. This highly social and gregarious nature is shared with great apes and monkeys. But dietary and life history changes in the Late Pliocene set ancient hominins on a separate evolutionary path that called for greater complexity in the structure and function of social networks. Social bonds were extended from interpersonal to intergenerational, from natal kin to cross-kin, and from local to regional. As noted above, these new social challenges were met by morphological and neurological changes in the brain that facilitated cognitive rather than impulsive behavior and cooperative rather than competitive zero-sum relationships.

Given these changes, it is likely that Pleistocene hominin groups were structured differently than those of contemporary apes. But were these ancient social groups large or small? Cohesive or atomistic? Uniform or diverse? As discussed in preceding chapters, the dominant portrait of ancient hominins is one of small, mobile groups of related individuals managing a subsistence-level existence in resource-limited habitats. The presumption of small group size and lack of social complexity is variously attributed to ecological challenges or to the limits imposed on sociality by the assumed pace of primate brain evolution.

One school of thought, represented by the hunting hypothesis, bases its conclusions on traditional notions about where hominins lived and what they ate. As we have seen, a prevailing view is that primeval habitats consisted of open savanna grasslands and that early humans were primarily dependent on the taking of large game. The limited carrying capacity of such habitats thus dictated low population densities and small community size. Boaz (1993: 227), for example, concluded that population density for early hominins was only 1.0–0.1 persons per square kilometer, or numbers roughly equivalent to that of contemporary South African Bushmen. This estimate was derived by comparing the ratio of fossil hominins to mammalian herbivores excavated at the Late Pliocene Omo site (Ethiopia), and then applying these same ratios to modern herbivore densities to arrive at proportionate ancient hominin population numbers. In a later publication, Boaz (1997: 160) reached similar conclusions about the low densities of *Homo habilis* groups. In this instance, diet assumptions and estimates of body size were utilized to conclude that available food resources would have limited these early hominins to small groups of fifteen to twenty individuals traveling over a broad subsistence range of about 1,000 square kilometers.

Ancient hominin density estimates by Boaz based on extrapolations from excavated fossil remains rely on two critical assumptions, both of which are arguable: (1) that the ratios of fossils recovered at the Omo site are representative of the actual densities of their respective taxa in ancient times, and (2) that the niches and densities of modern herbivores sufficiently mirror those of their ancient counterparts to be good predictors of ancient hominin populations. Herbivores and hominins may have lived in different niches within the same habitat, each of which has changed over the millennia. If ancient herbivores were savanna-adapted, but their hominin contemporaries relied primarily on the lacustrine food chain, then their comparative density ratios may be meaningless. Similar cautions apply to the *H. habilis* density estimates. Boaz assumes that, since early humans had a relatively large body size but a diet reliant on sparsely distributed resources, only small groups covering large ranges could be sustained. The reverse argument, however, also holds true. Namely, if dietary staples were abundant and were distributed in discrete but concentrated or resource-rich patches, then greater density and more stable communities could be supported.

Anthropologists have historically pointed to contemporary hunter-gatherers such as the Bushmen and Australian aborigines as models of early human society. Finlayson (2014: 142) cautions against this notion, noting the influence of ecological factors on demographic patterns. For example, although Australian desert foragers mirror the Bushmen in their small group sizes and large subsistence ranges, related aboriginal peoples formerly occupying the resource-rich banks of Australia's Murray and Darling rivers achieved population densities twenty to forty times that of nonriver peoples in the same region.

It is theoretically possible that such multipliers could also apply to more ancient hominins occupying resource-rich niches. As pointed out by Nicholas (1998), the wetlands habitats so characteristic of Pleistocene occupations played a role in both land use and population size due to the diversity, productivity, and reliability of the resources they supported. Pleistocene population density estimates err in assuming that ancient humans were always widely distributed over large home ranges in pursuit of game animals, as opposed to congregating at hot spots or subsistence hubs, either on a seasonal or year-round basis. Indeed, the alternating pattern of aggregation and dispersal in tandem with resource availability cycles is common among historic hunter-gatherers and may have been characteristic of ancient peoples as well.[14] Thus while, by the numbers, the total population of Pleistocene peoples may have been low by modern or even Neolithic standards, the relative density of their communities may have varied considerably on a seasonal or regional basis.

A second school of thought supporting the limited size and complexity of early social groups relies on the theory that hominin brain evolution proceeded at an attenuated pace. As noted in the discussion of the social brain hypothesis above, Dunbar proposed that the neocortex region of primate brains has a limiting effect on the ability to maintain social networks. Based on his study findings correlating primate neocortex size with group size, he concluded that the principal mechanism available to primates for dealing with the complexities of social group living was to establish and maintain intragroup bonds through social grooming. Dunbar calculated that the upper limit of time spent in grooming by primates in the wild, namely, 20 percent, could have been potentially extended to 30 percent in ancient times by social bonding interactions such as "vocal grooming" (i.e., distance calls). But beyond this threshold, the intensity of social grooming required for maintenance of more complex social bonds would be so time-consuming that corresponding increases in group size could not occur in the absence of language. Applying these assumptions to fossil neocortex estimates, Dunbar and Aiello (Aiello and Dunbar 1993) concluded that the group size of hominins was essentially mired at a level comparable to that of contemporary great apes throughout much of the Pleistocene. As noted earlier, Dunbar linked neocortex size not only to group size, but to the capacity for complex language, culture, and religion, which he argued are exclusive to *Homo sapiens sapiens*. Similar conclusions were reached by Gamble (1998: 443), who reasoned that the requisite expansion of hominin "social landscapes" did not occur until 100–60 ka. Barnard's (2011) three-stage theory for the co-evolution of language, kinship, and society is based on the same premise, and will be considered at greater length in chapter 6.

Dunbar is perhaps most widely known for his studies of group size among historic and modern populations. In what became known as "Dunbar's Number," he concluded that 150 represented the optimal number of relationships that can be effectively maintained due to natural limitations set by the human neocortex.[15] Subsequent studies have noted the tendency of human societies to organize themselves into social networks that, when expanded, create multilayered structures that increase in size according to a consistent scale. Zhou et al. (2005) calculated that a fractal-like scaling ratio of ~3 is applicable to hierarchical social network layers among modern human societies, while Hamilton et al. (2007a, 2007b) set the ratio for hunter-gatherers at ~3.8. This seemingly universal feature of human social organization may have some underlying biological basis. What is in question here, however, is not the existence of social scaling. Rather, it is whether inferences on the social scaling levels of Pleistocene hominins can be made based on the volumetric estimates of their brains.

Three principal arguments for early hominin small-group size posed by the social brain theory have been recently challenged. First and foremost are neurological study findings cited above that discount previous assumptions about disproportionate neocortex growth. New research not only documents that primate brains were subject to linear scaling over time, but suggests that the critical factor for "social intelligence" in brain evolution was neurobehavioral specialization. Similar-sized brains may have dissimilar patterns of neuronal allocation and functional interconnectivity. In short, without the ability to scientifically probe the anatomy of gray and white matter in ancient hominid brains, there is no reliable way to determine what may have actually been going on inside.

A second consideration is the social metric on which the social brain theory is based. While the "you scratch my back, I'll scratch yours" concept may be a valid yardstick with which to measure social bonding investments among contemporary apes, one may well question the math of applying this back-scratching calculus of time management to Pleistocene hominin communities. Other types of reciprocal acts for forging social networks could have been equally effective and perhaps less time-consuming, particularly when the potential for an early evolution of the language instinct is considered. For example, de Ruiter, Weston, and Lyon (2011) argue that the establishment of relationships and the tracking of interpersonal transactions are enabled by cultural means that allow individuals to move beyond alleged neocortical limitations such as those proposed by Dunbar. The authors conclude that, given the potential variety of cultural inventions for social bonding that may evolve in the absence of genomic change, neocortical size alone is a poor predictor of group size and complexity. Social learning provides cognitive shortcuts and novel means for the tracking of reciprocal transactions essential to group living.

Finally, the social brain model fails to integrate ecological and evolving behavioral plasticity factors into the equation of complex social bonding and group size. Dunbar specifically rejects the conclusions of Reader and Laland (2002), discussed earlier, that behavioral plasticity, social learning, and ecological challenges played the predominant role in brain evolution and social group characteristics. In so doing, he constructs a false dichotomy between "ecological" versus "social" problem-solving. Says Dunbar:

> It is important to appreciate in this context that the contrast between the social and more traditional ecological/technological hypotheses is not a question of whether or not ecology influences behavior, but rather is one of whether ecological/survival problems are solved explicitly by individuals acting on their own or by individuals effecting social (e.g. co-operative) solutions to these problems. In both cases, the driving force of selection derives from ecology, but the solution (the animal's response to the problem) arises from contrasting sources with very different cognitive demands (individual skills in one case, social-cognitive skills in the other). (2003: 164)

The false dichotomy rests in the fact that ecological and social problem solving in early hominin groups were *wed*—both involved the enhancement of social-cognitive skills for the creation and maintenance of cooperative strategies. The essence of eusociality is a division of labor in which the members of a social group engage in reciprocal altruism. Since, for female hominins ("energy maximizers"), social networks are structured in relation to the food supply, all problem solving directly or indirectly involved the development of cooperative solutions to ecological/survival issues. Female-bonded groups were, in essence, corporate regimes with a basis in energy exchange. The ecological solution to problems of enhanced food access was cooperative foraging and food sharing, which relied on the creation of alliance networks, often multigenerational in scope, for resource allocation and distribution. Similarly, the evolution of alloparenting and cooperative breeding systems was a corporate solution to ecological/survival problems that required the establishment of intergenerational alliance networks. In each case, social-cognitive skills for enlisting group participation were integral to problem solving. And in each case, the positive feedback mechanisms used to bind individuals and social groups were both tangible (food commodity exchange) and intangible (energy savings).

Contemporary primate groups vary in size from one to several hundred individuals depending on resource availability. Baboon troops, for example, range in size from 5 to 250 individuals, and geladas from 30 to 350, with increases to over 650 when food is abundant. Moreover, members of the same or closely related taxa with comparable neocortex volumes exhibit different group sizes and social characteristics in response to changing environmental conditions. This same potential for diversity is likely to have existed for ancient hominins. The role of behavioral plasticity and ecological problem solving in shaping the size and structure of primate communities must therefore be integrated into theories about the nature and size of Paleolithic hominin groups. The essential question is: Did members of early *Homo* pursue their reproductive strategies as individuals in small, isolated bands? Or did the process of becoming human involve, or indeed require, the forging of cooperative strategies in the context of larger multimale-multifemale communities?

Prevailing ecological models have focused on the relative costs and benefits of animals grouping into large communities. While strength-in-numbers advantages such as reduced predation risk and increased defensive capabilities are frequently cited as beneficial outcomes, aggregation also places a heavier burden on the sustainability of local resources. As group size increases, greater feeding ranges and expenditures of time and energy are required to support their memberships.

The Ecological Constraints Model proposed by Chapman and Chapman (2000) holds that group size is a function of the size, density, and distribution of food resources. Where resource patches are of high quality, are densely distributed, and involve low travel time, large group sizes can be sustained. In contrast, when food resources are of low density and uniformly distributed, only small groups can be supported due to feeding competition, travel costs, and resource depletion.

Changing ecological factors may also influence the size and composition of social groups over time. Several species of primates, including humans, have adopted cyclical fission-fusion patterns, wherein a parent population alternatively disperses and aggregates in response to seasonal or stochastic variation in the nature and distribution of resources.[16] Such ecological problem solving would have been particularly valuable to hominins living in the mosaic habitats and volatile climate conditions of the Plio-Pleistocene. The cyclical concentration of group members has also been linked to the ability of evolving hominins to move between time and space—to create social networks founded on common kinship and to create and sustain multilevel political communities. On this basis, Gamble (2008: 40) proposes that there may have been strong selection for early hominins to occupy habitats that allowed for periodic fusion and the opportunities such gatherings provided for intense social interactions.

The ecological constraints model is particularly relevant for an understanding of demographic patterns throughout the Pleistocene. Finlayson (2014: 128–31), in his survey of Upper Paleolithic sites, makes a distinction between "source" and "sink" populations. Source populations are those that generate a surplus of members that emigrate from their native communities. These populations are commonly associated with sites located in proximity to water sources and the coast. Sink populations, in contrast, are found in more marginal habitats and operate at a demographic deficit due to resource stress or scarcity.

Evidence suggests that Pleistocene populations lived in communities of varying size, density, and complexity depending on local conditions. Those living in biodiverse habitats were most likely to experience population growth and to eventually colonize contiguous areas. The successive dispersals of various hominin lineages during the Pleistocene and Holocene periods are perhaps best explained by this dynamic. These colonizations were not the result of purposive crusades by wandering hunting bands, but rather by the incremental dispersal of overflow populations from subsistence hubs or hot spots whose carrying capacities had been approached or met. Such land use, settlement, and dispersal patterns may be as old as our genus.[17]

It is interesting to view the Plio-Pleistocene origins and spread of *H. erectus* in this light. Stewart's (1994) analysis of East African early hominin diet and subsistence concludes that reliance on aquatic fauna such as fish had important sociodemographic implications. The abundance of these easily acquired, high-quality foods could have facilitated greater sedentism and the concentration of larger groups living in close proximity to lakes and rivers. At such hot spots, members of waterside encampments would have aggregated and dispersed in cadence with the seasons and the relative abundance of staple foods. Although fish and shellfish would be available on a year-round basis, the largest gatherings would be expected to coincide with annual fish spawning cycles. As among contemporary hunter-gatherers, such seasonal aggregations would provide opportunities for sociality, mating, and the forging of social alliances. *H. erectus* communities organized in this way would have been well-nourished, highly social, and reproductively successful.

The combination of abundant high-quality foods and cooperative breeding would have set the stage for rapid population growth. *H. erectus* populations in densely occupied riverine and lakeshore niches would be expected to periodically fragment in response to tectonic events and climatic oscillations that impacted the distribution of water and aquatic fauna. Demographic pressures and resource shortages would have motivated *H. erectus* communities to evacuate and disperse. When they did so, it is likely that they followed major watercourses to more northerly refugia and to the coasts.

Available evidence suggests that *H. erectus* population movements throughout Africa and across the broad swath of Eurasian middle latitude belts occurred in waves over thousands of years in response to such demographic pressures and climate oscillations. The ecological release of *H. erectus* into new but compatible habitats was accelerated by the lack of competitors in their path. As they moved, they likely selected occupation areas that were most familiar to them, namely, well-watered, biodiverse habitats. River valleys and the shorelines of both the Mediterranean Sea and Indian Ocean provided likely migration corridors for peoples already adapted to the subsistence gathering of marine and freshwater foods.

This model of hominin land use and settlement patterns explains how *H. erectus* as well as subsequent Neanderthal and AMH populations expanded over vast regions of the Old World—by aggregating at subsistence hubs, multiplying, fragmenting, and then dispersing into new contiguous habitats when the carrying capacity of their natal area was reached. In searching for contemporary examples of what Paleolithic communities may have been like, therefore, it is important to look to both source and sink populations. For example, Australian aborigines living on the Murray and Darling rivers, the prehorse Okanagan of the North American Columbia River Valley, Native Americans of the California coast, or the pedestrian hunters of South America's Gran Chaco may provide valuable insights into the variable densities and cyclical aggregation patterns of ancient hunter-gatherer communities. Observations of Australian desert bands and !Kung Bushmen are valuable in that they more closely mirror sink populations that struggled to weather the uncertainties and climate extremes of the Pleistocene. They should not be taken, however, as prototypical of ancient peoples that had unfettered access to resource-rich habitats and the ability to congregate into multilevel communities. The unifying characteristic of hominin social groups, it seems, is that they are inherently drawn to one another, to the formation of cooperative socioeconomic groups, and to membership in the largest communities that resources allow.

The tendency of hominins to pursue survival strategies in groups sized to the maximum allowable limits of carrying capacity in a given niche is prominent among some nonhuman primates as well. Notable in this regard are those occupying terrestrial habitats where multimale-multifemale groups, intra- and intersexual alliances, and large group size provide economic and reproductive advantages. The alternative scenario for hominin evolution presented in chapter 3 proposes that this type of sociosexual organization would have provided the necessary energy budget and personnel for development of both broad-spectrum diets and cooperative breeding communities in the Late Pliocene. When combined with a well-watered and biodiverse environmental setting, this model for early human groups would lead to different land use patterns and greater community size than are proposed by prevailing theories.

Energetics and Labor Division

One of the characteristics that defines our genus is the organization of individuals into groups whose members are bound by cooperative socioeconomic relationships. In hominin groups, reciprocal alliances structure both the allocation of work and the distribution of resources among members, in effect managing the "energy budget" of the collective.

Discussions of labor and material flows in ancient society have traditionally focused on the differentiation of male and female subsistence roles. These theories, interestingly, address the earliest and the latest periods of the Pleistocene. As noted in chapter 2, advocates of the hunting hypothesis propose that the division of labor by sex arose as a natural consequence of pair-bonding relationships at the dawn of humanity. More recently, Kuhn and Stiner (2006) have theorized that subsistence roles were relatively undifferentiated by sex until late in human history when AMHs introduced a shift to broad-spectrum diets. Neither of these proposals has been supported here, in the first instance due to the lack of evidence for early male provisioning, and in the second because wide dietary breadth is also characteristic of much earlier hominins.

An alternative approach to understanding ancient socioeconomic life, and one more aligned with this book, is the investigation of production and distribution activities of social group members from the perspective of energy exchange. Energetics is the study of energy flows and storages under transformation. As discussed in chapter 3, its principles have been applied by several scholars to highlight critical shifts in energy input and distribution that marked the first division of labor in evolving human groups, namely, the emergence of cooperative breeding (Foley and Lee 1991; Aiello and Key 2002; Isler and van Schaik 2012). To reiterate, females offset the increased reproductive energy burden accompanying hominin encephalization with two effective strategies: expanding the quality, quantity, and variety of foods procured; and creating a network for reciprocal allocare and food sharing. Over time, the adaptive advantages of these behaviors were realized through an improved food supply, reduced mortality rates, increased fecundity, and shorter birth intervals. As pointed out by Isler and van Schaik, without this fundamental redistribution of energy budgets toward mothers and infants that sustained population growth, the process of hominin brain expansion would have hit the "gray ceiling" shortly after it began. This premise on critical energy reallocation is also supported from a neurological perspective by Herculano-Houzel et al.:

> Thus, we reinforce the previous conclusion that the human brain is not extraordinary within primates, given that the same scaling rules apply to it—although it is remarkable in the total number of neurons that it contains, superior to that in all other primates, possibly due to the overcoming by human ancestors of the energetic limitations that presumably curb further increases in numbers of brain neurons in non-human primates beyond what is found in extant great apes. (2014: 26)

Viewed through the lens of energetics, all forms of labor division in hominin societies can be understood as specializations of social group structure and function—strategies for maintaining and improving energy flow and efficiency. In the case of alloparenting, labor division was accomplished through the establishment of sterile classes, composed of postmenopausal and premenstrual females, whose collaborative efforts enhanced the survival of altricial young while freeing fertile female group members to pursue a broader spectrum of activities. Other types of labor division, such as the differential allocation of work based on age, skills, knowledge, strength, mobility, and other individual attributes, is also expected to have occurred in early human societies as a natural consequence of group socioeconomic activity. Foraging strategies that optimized labor for essential food-getting activities would have been selected for based on their greater productive yields. These strategies would be expected to vary with the type of resource being procured and nature of the available workforce. The division of labor by sex is not preordained, but may have appeared in hominin societies throughout the Pleistocene whenever the de facto attributes of resident males and females impacted productive efficiencies.

The specialization of structure and function also had a critical spatiotemporal dimension in ancient society. Resources were unevenly distributed not only geographically, but often seasonally as well. The life cycles of some taxa, such as fish spawning runs or the predictable migrations of terrestrial mammals, presented intermittent opportunities for enhanced resource procurement and storage that could be optimized by large work forces. As among historic hunter-gatherers, communal rabbit drives or the stampeding of herd animals into ambush likely required the assembly of all able-bodied men, women, and children from one or even several communities. This division of labor thus involved not only sex and age considerations, but the coordination of local or regional socioeconomic units as well. In this and in other types of Pleistocene communal subsistence activity, humans adapted to fluctuating patterns of resource availability by developing integrative social networks that mobilized large labor forces and maximized their collective ability to capture and transform energy.

Ludwig von Bertalanffy, in his seminal work *General Systems Theory* (1968), formulated elemental organizing principles and processes and applied them to all organic forms. This metatheory was based on the assumption that all life forms represent "systems" that rely on the successful transformation of energy in the environment for their sustenance, reproduction, and survival. Systems and their component parts are sustained by resources, maintained via some technology, and impacted or shaped by physical and social externalities. Systems undergo life cycles comparable to living organisms, the rhythm of which is dictated by adaptive successes and failures.

This model translates rather readily to both ancient and modern societies, in which corporate socioeconomic groups operate as systems. In a society's dynamic energy exchange with its environment, fitness-related resources are imported (*inputs*), transformed or processed (*thruputs*), and exported (*outputs*). Typical inputs may include various sources of energy, natural resources or materials, or other productive factors, along with genes and less tangible resources such as information, ideas, and symbols. Thruputs include the various technologies employed to process or transform energy and resources into outputs, such as material and symbolic products and reproductive fitness that sustain the system (see Figure 5.2).

Thruput is a dynamic process wherein activities and social networks are defined and redefined to meeting changing system needs. Systems are aided in their quest for survival by two important processes. First, *differentiation* provides for specialization of structure and function to enhance communication with the environment and to better accomplish the energy exchange process of a population in a given niche. These social structures are the vehicles through which energy transformation takes place and represent collections or coalitions of individuals or interest groups that pursue various thruput strategies on a collaborative, competitive, or symbolic basis. The greater the degree of internal differentiation and heterogeneity, the greater the number of "technologies" available to the system to both access and process resources.

Second, systems receive *feedback*, that is, positive and negative signals from the environment that provide a yardstick by which to measure the appropriateness or success of behaviors. Feedback is the system's information network whereby messages regarding interaction with the environment are relayed back, interpreted, and assessed. Messages include empirical or tangible data on the nature of the external environment and the system's relationship with external groups and things, as well as data that define the boundaries of uncertainty for the system and its component groups and that provide the symbolic building blocks for myth, ritual, and ceremony. This critical exchange promotes *negative entropy* (the absence of entropy) and *homeostatic equilibrium* through a continuing process of adaptation and through the reintegration of those parts of the system that are central to the necessary and efficient transfer of energy.

To the extent that systems actively pursue stability and survival, they are regarded as purposive. This goal-directed action or notion of causality, however, is not linear and should not be confused with "progress." Rather, von Bertalanffy invents the term *equifinality*, a kind

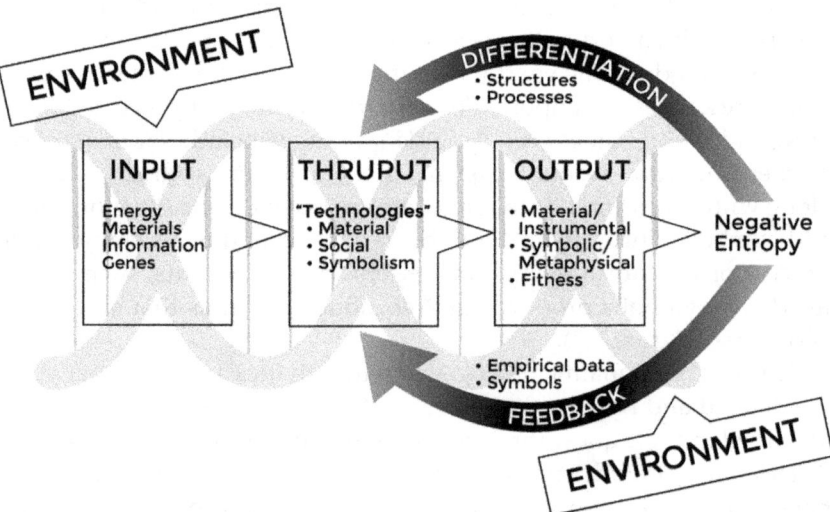

Figure 5.2. A model of human society based on general systems theory. By Drew Fagan.

of multilinear causality, to distinguish the tendency of systems to accomplish a desired end state through a myriad of starting points and by an unpredictable variety of paths. Notably, since externalities are not static, the idealized state of equilibrium is elusive and systems must remain in constant flux to maintain negative entropy.

According to general systems theory, the more differentiation that occurs within a system, such as specialized structures for the organization of labor, the more "technologies" a system has at its disposal for energy exchange.[18] Paleoanthropologists have traditionally equated Pleistocene technologies with material culture, and specifically with lithic traditions. Less often has the discussion moved from hardware to "software." Oldowan and Acheulean tool inventories dominated the first 1.5 million years of human existence. If encephalization and biocultural evolution during this period relied on an increasing mastery of resource exploitation and energy exchange, how can human advancements be explained in the absence of obvious technological innovation? It has been proposed here that the critical technologies invented and applied by *H. erectus* and descendant hominins such as *H. heidelbergensis* and *H. rhodesiensis* consisted of social networks by which their communities were organized to achieve greater efficiency in energy extraction and exchange. In other words, early hominins developed *social* technologies and symbolism to augment their limited material toolkits in meeting the challenges of dynamic environments. Social technologies evolved to optimize the procurement and distribution of "fitness-related resources" such as energy, materials, genes, and nongenetic information to group members (Hamilton et al. 2007a, 2007b).

During the latter half of the Middle and Late Pleistocene periods, toolkit refinements such as stone blades and projectiles appeared, greatly enhancing the range of technologies available to ancient peoples for capturing energy inputs. New and more effective instruments for the taking of large game, such as Mousterian tools, allowed for parallel adjustments in social technologies. What formerly may have required the collective efforts of all community members for the taking of large herbivores could now be effectively achieved utilizing a much smaller group of skilled and newly weaponized hunters, thus freeing other segments of the work force to pursue alternative economic endeavors. The later appearance of microlithic technologies greatly accelerated this trend, increasing overall efficiencies for accessing protein energy inputs while reducing the percentage of the population directly engaged in their acquisition. The end result was a dramatic increase in productivity and labor specialization and, over time, the accumulation of surpluses that laid the foundation for open economies and sedentary life. The emergence of Late Pleistocene peoples and their transition into the Neolithic Age, therefore, need not rest on notions of cognitive exceptionalism. They simply built on the social and material technologies invented by their ancestors to enhance system energetics, a process that continues to the present day.

Social technologies were constructed through the idiom of kinship for most of human history. Groups based on common kinship formed the backbone of Paleolithic societies, structuring every facet of economic and reproductive life. It was in mother-offspring-sibling units that a critical shift in group energetics first occurred. Cooperative breeding communities and eusociality arose out of the division of labor among close kin for food getting, food sharing, and allocare. In Paleolithic society, differentiation of structure and function for energy exchange was built on the foundation of consanguineal kinship groups that not only allocated work, but regulated the procurement, processing, and distribution of resources. These units of cooperating kinfolk operated as systems, optimizing their social technologies to adapt to changing environmental conditions. They also formed the building blocks of larger systems or collectives, the component groups of which may have aggregated and dispersed on a seasonal basis in cadence with resource availability. Differentiation within these composite communities consisted of complex alliances among kinship groups for collaborative activity that enhanced the energy budgets of all participants.

Consanguineal kinship groups provided an efficient organizational framework not only for the exchange of energy and materials, but for the exchange of genes as well. The founding principles of these groups and their role in shaping the course of human evolution is perhaps one of the most studied and most debated topics in contemporary anthropology. The final two chapters of this book explore past and present theories on the origins of male and female philopatry and how the concept of social DNA sheds light on matrilineal and patrilineal kinship organization as alternative strategies for niche construction.

 6

Kinship and Paleolithic Legends

Mark Twain was once quoted as saying that sacred cows make the best hamburger. Theories on the origins of human kinship systems are replete with sacred cows, but picking them out of the herd, let alone rendering them, has been a challenging enterprise. Metaphors aside, theories about the determinants, characteristics, and phylogeny of male and female philopatry and unilineal descent systems have probably engendered more extended and heated debate than any other topic in anthropology. A primary reason is that this subject has historically been caught in the crosshairs of science, politics, and religion. When it comes to questions about the nature of human nature, sexuality, primitive communism, the relative dominance of the sexes, or the origins of social inequality, scholars have inevitably brought their cultural values to the table. And, as frequently happens in philosophical debates, arguments tend to be formulated intuitively and promulgated on the basis of what appears, through the cultural mind's eye, as the natural order of things. Because such theories are often not attestable or, worse, are impervious to facts, they have become anthropology's Paleolithic legends—zombie ideas that arguably have enjoyed undeserved immortality in the discipline. Some of these concepts still claim vocal advocates and have, in this writer's opinion, hindered progress on our understanding of ancient and contemporary social groups. Our discussion of kinship thus begins by examining some of academia's persistent biases about the evolution of human social organization and how these ideas have contributed to the status of current theory. It also explores why genomic research may represent a new frontier in reconstructing the origins of human kinship systems.

Universal Evolutionary Stages

The earliest and most persistent types of origin theories propose that human evolution has proceeded through a discrete set of universal stages. Inherent in all such schemes is the notion of progress, if not by some remote design, then simply by a trending from primitive to more advanced. These macro-theories have enjoyed great popularity over the past 150 years and in each case have had something important to say about human kinship.

Some of the earliest examples are found in the writings of nineteenth-century evolutionists, such as Marx, Bachofen, McClennan, Morgan, Lubbock, Spencer, and Engels, who generally saw human society as progressing from a state of primitive promiscuity, to mother-right or matriliny, and ultimately to father-right or patriliny.[1] Implicit in these theories was the notion that ancient peoples emerged from sexual anarchy and a general ignorance of biological reproduction to a recognition of parental relationships, the maternal being more obvious and primitive, while the paternal represented an assertion of male dominance and more refined, symbolic links to offspring.

As noted previously, such schemes were enthusiastically dismissed in the early twentieth century (at least in Western science), along with unilinear evolutionism generally. When the so-called neoevolutionist movement emerged in the early 1950s, greater emphasis was placed on drawing parallels or regularities from the expanding ethnographic record and on recognizing the influence of ecology on cultural adaptations and culture change. The architects of new evolutionary schemes, such as Julian Steward (1955) and Elman Service (1962), described their frameworks as *multi*linear rather than unilinear and instead proposed general categories of sociocultural complexity. While these categories fell into a logical evolutionary sequence (i.e., the band, tribe, chiefdom, and state), they were not represented as universal stages in the sense that all of humanity passed through them. The obvious exception was the band level, which was presumed to characterize ancient as well as contemporary hunter-gatherers.

Notably, neoevolutionist assumptions about kinship at this earliest level of human society took an abrupt about-face from prior theory. Steward argued that one of the principal inadequacies of nineteenth-century models was their assignment of evolutionary priority to "matriarchal" kinship systems. By a contrary and somewhat convoluted argument, he concluded that the simplest level of societal complexity, namely, the "band," was necessarily patrilineal:

> We now have to consider why these bands are patrilineal. First, it is characteristic of hunters in regions of sparse population for post-marital residence to be patrilocal. This has several causes. If human beings could be conceived stripped of culture, it is not unreasonable to suppose that innate male dominance would give men a commanding position. If, in addition to native dominance, however, the position of the male is strengthened by his greater economic importance, as in a hunting cul-

ture, or even if women are given greater economic importance than men, it is extremely probable that postmarital residence will be patrilocal. (1955: 125)

The priority of male-centered kinship groups at the simplest level of sociocultural complexity was also adopted on similar grounds by Service, who stated categorically: "There are no purely uxorilocal-matrilocal band societies" (1962: 60). As noted in chapter 2, twenty-first-century evolutionary reconstructions, such as those proposed by Rodseth et al. (1991), Jolly (2009), and Chapais (2014), have followed suit. These sociobiological models commonly place reliance on innate male dominance and male inclusive fitness as the principal drivers of early human social life and also credit male philopatry with the development of multilevel social groups and more complex levels of societal integration.

The authors of these theories would undoubtedly be quick to dissociate themselves from their nineteenth-century forebears. But there are at least two important ways in which their theories on the nature of early human society are similar and that have collectively skewed our perception of human kinship. First, all make summary judgments about the appropriateness of kinship types based on the relative antiquity or sociocultural complexity of ancient peoples. For example, most conclude that kinship reckoning through females among the earliest or simplest human groups was either a transitory and primitive form, or was entirely absent. Postmarital residence and descent in early society are presumed to have been necessarily patricentric due to the alleged sexual dominance and fitness strategies of males. Moreover, male-centered kinship is frequently characterized as a more stable or advanced social form than its female-centered counterpart at all levels of sociocultural complexity.

Second, these theories have consistently tied prevailing kinship patterns to specific subsistence technologies, and in particular to the presumed degree of sex-biased patterns of cooperative activity. Thus, hunter-gatherers have been historically characterized as naturally patrilineal and patrilocal due to assumptions about the importance of male bonding in the hunt. A similar logic, namely, cooperative farming activity by women, has been utilized to explain not only the presence but the alleged limitation of matriliny to horticultural societies. As will be discussed in the next chapter, such assumptions are negated by the ethnographic record—by the documented presence of viable matrilineal-matrilocal hunter-gatherer societies and by both matrilineal and patrilineal alternatives among horticulturalists, which vary independently with sex-biased subsistence contributions and patterns of cooperation.

When discussing matrilineal and patrilineal kinship in the Paleolithic, it is of paramount importance to note the conditions under which each type of social organization may have been adaptive. There may well turn out to be recognizable patterns in the frequencies of these organizational options over time. But such trends must be independently linked to the underlying conditions that favor specific social technologies rather than to preconceived notions about human nature or an inherent direction for gene-culture co-evolution.

The notion that human social origins and early systems of kinship can be explained on biological or ecological grounds has generally been regarded as an anathema by British social anthropologists. Scholars devoted to ethnography and synchronic studies have been openly critical of such models and have recently demonstrated renewed interest in joining the conversation on how both kinship and language emerged in evolving hominins (Allen et al. 2008; Power et al. 2017). They argue that a nuanced interdisciplinary perspective is needed that combines data from paleontology, archaeology, primatology, and evolutionary psychology with insights drawn from the ethnographic record. Resultant models have proposed universal stages of cognitive development that chronicle the co-evolution of kinship, language, and symbolic systems.

Prominent among these is the "new synthesis" model formulated by Barnard (2008, 2009, 2011), which builds on Dunbar's (1993, 2003; Aiello and Dunbar 1993) social brain theory (the co-evolution of neocortex size and group size) and a three-stage theory of language evolution proposed by Calvin and Bickerton (2000). Barnard suggests that the evolution of human society proceeded through three distinct stages, designated as the Signifying, Syntactic, and Symbolic Revolutions, to which are assigned representative hominin forms and characteristic levels of group size, symbolic communication, and kinship complexity. Briefly, the Signifying stage is represented by *Homo habilis*, living in small groups with only proto-language (gestures, words), "mate sharing," and Hawaiian kinship terminology (merger of siblings with cousins). The Syntactic Revolution, associated with Archaic *Homo sapiens* (i.e., *H. heidelbergensis* and Neanderthals), is marked by larger group sizes and intergroup exchanges, rudimentary language (sentences and simple syntax), and kin exogamy. The third or Symbolic Revolution is reserved for the appearance of modern *Homo sapiens*, at which time group sizes increase to Dunbar's optimal 150 number, along with the emergence of reciprocal exchange systems, "true" language (full syntax), and "true kinship" (universal kin categorization).

In outlining the characteristics of the three stages, Barnard makes reference to early universal evolutionary theories—i.e., the similarity of the Signifying stage with Morgan's primitive promiscuity, and the congruence of McClelland's theory on the evolution of exogamy as a marker of the Syntactic Revolution. Barnard places little emphasis on the question of uterine or agnatic priority in kinship reckoning, although he clearly sides with Isaac (1978) on the central importance of hunting, pair bonding, and parental investments by males for hominin social advancement. The principal focus of his three-stage model remains on the perceived consequences of neocortex expansion for symbolic communication and its expression in increasingly complex linguistic and kinship structural elements.

Ultimately, then, the strengths or weaknesses of this three-stage model lie in the predictive value of Dunbar's social brain theory itself, which Barnard accepts as one of the "unconscious truths" in human evolution (2011: 131). If, however, as recent neurological studies suggest, the cognitive abilities of ancient hominins cannot be reliably discerned by the relative size of their forebrains, then there is no material basis on which to assume that *Homo erectus* could only speak simple words, or that *Homo neanderthalensis* was cognitively incapable of enjoying fully developed kinship and mythological systems.

Kinship Theory and Politics

As a bright-eyed anthropology graduate student in the 1960s, I once approached a senior professor known for his vast in-office library for a book loan. After requesting a copy of Frederick Engels's *Origin of the Family, Private Property, and the State*, I was quickly hustled behind a closed door and informed, in hushed tones, that such books were "not discussed," and were certainly not worth reading. It was only later that I became aware of the persistent emotional fallout affecting some academicians who had lived through the 1950s McCarthy-era purge and, in a larger sense, the extent to which politics has influenced the development and direction of scholarly endeavor. The discussion of zombie ideas, therefore, would be incomplete without reference to Morgan's pioneering works on human kinship, the incorporation of his observations by Marx and Engels into the framework of dialectical materialism and Communist political theory, and the summary rejection of these ideological premises by the emerging American and British schools of anthropology. Harris (1968), in his comprehensive tome *The Rise of Anthropological Theory*, observed:

> With Morgan's scheme incorporated into Communist doctrine, the struggling science of anthropology crossed the threshold of the twentieth cen-

tury with a clear mandate for its own survival and well-being: expose Morgan's scheme and destroy the method on which it was based. The anthropological attack against Morgan was to have these consequences: (1) the abandonment of the comparative method; (2) the rejection of the attempt to view history from a nomothetic standpoint; (3) postponement of actual tests of the cultural-materialist strategy during a forty-year interlude. (1968: 249)

Indeed, Harris argued that American and British anthropology developed in *reaction* to Morgan and Marxism. Founding fathers of the discipline in the United States, such as Boas, Kroeber, and Lowie, were quick to dismiss the concept of evolutionary stages, along with the notions of ancient promiscuity, primitive communism, and the priority of matrilineal descent.[2] The basic premise of the Bachofen/Morgan matriarchate, namely, literal rule by women, was especially unpalatable to Western scholars, conjuring up as it did cultural images of Amazons, henpecked husbands, and overbearing mothers-in-law. The notion of ancient gynocracies was rather easily dismissed by virtue of their absence in the ethnographic record. A generation of American anthropologists was further instructed on the futility of identifying cultural uniformities or laws and was advised to instead pursue a more descriptive, ideographic approach to ethnology.

Likewise in Great Britain, Westermarck (1894, orig. 1891) was an early and vocal critic of Morgan's ideas on primeval mating patterns, which British anthropologist Meyer Fortes later disdainfully dismissed as the "repugnant hypothesis of promiscuity and group marriage" (1969: 5). Founders of the British structuralist-functionalist schools also soundly rejected Morgan's concept of the ancient matriarchate and instead devoted decades to the promotion of synchronic kinship analyses.[3] Thus, with two of the three foundational pillars in Morgan's evolutionary scheme effectively removed, the stage was set in Western anthropology by the 1950s for recrafted theories on human origins based on male philopatry, pair bonding, and the sexual contract.

Meanwhile, in socialist countries, the discipline of anthropology developed in the opposite direction. Briffault's (1963, orig. 1927) enthusiastic defense of evolutionism and of Bachofen's original ancient matriarchate was widely accepted in emerging twentieth-century Soviet anthropology. There, an initial matrilineal epoch was generally regarded by scholars as a template for preclass society. Consistent with Engels's materialist refinement of Morgan's scheme, overthrow of the female-centered primitive collective by patriarchal organization and the nuclear family was proposed to have been initiated by a fundamental change in the nature of property relations that accompanied the emergence of social classes.

Given the divergent trajectories of Euro-American and Soviet anthropological theory, it is no surprise that academic exchanges occasionally became quite animated, with accusations of demagoguery hurled from both sides of the political divide. In the United States, Leslie White (1949, 1959) spearheaded a softening of ideas on evolutionism and materialism by the mid-twentieth-century mark, and increasing attention began to be paid to the relationship between economic adaptations and the nature of social forms. At the same time, Soviet anthropologists had begun to exchange the concept of matriarchy (or gynocracy) for matrilineal kinship and to abandon the notion of strict priority for matriliny over patriliny by recognizing the potential for their coexistence or independent origins.[4]

It would be naive to assume, however, that ideological purity in both camps has gone completely by the wayside. Some Soviet anthropologists continue to support the theory that initial kin collectives were exclusively matrilineal, although others concede that their replacement by patrilineal institutions may occur at any point in history where property relations are transformed from collective to individual ownership (Semenov 1965). Similarly, as has been noted throughout this book, there is a sustained bias in Western anthropology that the earliest forms of human social organization were necessarily male-centered. Indeed, the mere suggestion of ancient matriliny or discussion of Morgan's take on classificatory kinship systems still engenders pronounced eye-rolls in some academic circles, or even overt hostility in scholarly exchanges.[5] As one commentator put it: "The question of the matriarchate or matriliny in regard to the periodization of the social-cultural history of the human kind has remained with us for over a century as the unlaid ghost of anthropology" (Krader 1979: 350).

Knight (2008) argues that the politicization of theories on early human kinship led to a schism within the discipline and the subsequent reluctance of social anthropologists to participate in the dialogue on our evolutionary origins:

> Morgan's work on the matrilineal clan had led such influential thinkers as Engels, Freud, and Durkheim to argue for fundamental discontinuity between primate and human social organization. Classificatory kinship, exogamy, totemic avoidances—in any but the most blinkered account of human origins, such things simply cried out for explanation. But Morgan's suppression marginalized evolutionary questions and therefore sidelined social anthropology's distinctive scholarly contribution to evolutionary science. From this point on the two branches of anthropology were hardly on speaking terms. As a result, Darwinians became cut off from specialist knowledge about cross-cultural variability in human kinship arrangements and from processes driving historical change. Forced

to draw narrowly on their own cultural assumptions, would-be Darwinian scientists recurrently mistook monogamy, paternal inheritance, and other contemporary instantiations of Judaeo-Christian morality for core features of human nature. (2008: 74)

Over the past decade, there is evidence of a renewed interest in kinship theory generally and a willingness of social anthropologists to apply their insights and expertise to questions of human social origins. This promises not only to reopen old evolutionary debates, but to do so with the benefit of new interdisciplinary data.

In summary, it is important to acknowledge that politics has the ability to compromise objectivity and skew academic perspectives. When it comes to reconstructing the sociosexual life of Paleolithic peoples, scholars now have a vast universe of biological, paleontological, archaeological, and historical data on which to draw. It is in the selection and interpretation of these data—whether, for example, one takes an emic or an etic approach, or whether one chooses to follow the selfish gene or follow the money—that cultural bias comes into play. The remnants of Cold War anthropology persist in kinship theory, despite the fact that cultural anthropologists have been trained to recognize ethnocentrism when they see it and should probably know better.

Innate Male Dominance

Another approach to understanding the roots of human kinship systems focuses on the dimorphic characteristics of the sexes and their perceived implications for political and socioeconomic dominance. The basic theory is that males, by virtue of their biology, are naturally predisposed to exercise hegemony over females and therefore to monopolize the formation and management of kinship groups, the control of wealth, and the allocation of power and authority. In this model, the universal basis for male-centered kinship groups harkens back to the origins of society, in which male provisioning and control of female reproduction are credited with establishment of the human family.

The role of intersexual dominance in the formation of social groups is typically related to male physical prowess and disparate reproductive responsibilities. Males are generally bigger, stronger, more aggressive, and relatively emancipated from ongoing temporal commitments to infant care. Females, in contrast, are generally of slighter build, more hormonally attuned to nurturance, and carry a greater reproductive burden through serial pregnancies, parturition, and the care of others. Early comparative ethnologists such as Murdock (1937) correlated such dimorphic physical traits and reproductive roles with characteristic patterns in the division of labor by sex, noting that the majority of tasks assigned to women can be performed in close proximity to the household, whereas male task assignments associated with subsistence activities or warfare require greater stamina and take place at greater distances from the dwelling unit.

Such observations led Murdock to propose the nuclear family or male-female dyad as the universal socioeconomic unit, with female activities focused in the domestic sphere and males prevailing in the external or political domain. Murdock supported the argument for male-centered postmarital residence patterns since, he concluded, the subsistence activities of men require an intimate knowledge of the local terrain and resources that would be compromised if they were obligated to leave their natal territory (Murdock 1949: 213–14). Significantly, these asymmetrical sex-linked patterns were not seen as limited to specific evolutionary stages or levels of sociocultural integration, but rather as defining characteristics of the human condition across time and space. On the basis of dimorphic traits, it was concluded that the human default pattern called for nuclear families, male philopatry, and male dominance in the economic, social, and political domains. Accordingly, nonconforming institutions or societies must be understood as aberrations, or as highly specialized exceptions to the rule.

In the 1970s, a lively academic debate on the subject of innate male dominance took place, the major parties to which have since passed, or have retired to their neutral corners with original mindsets intact. The issues raised in this dialogue, however, remain substantially unresolved and are directly relevant to theories on the origins of human social forms. As we have seen, innate male dominance is a central tenet of both the ancestral male kin hypothesis and gene-based sociobiological theory. A central issue in this debate has been how to explain or understand the predominance of male-centered social institutions and ideologies of male superiority found in the current ethnographic record.

One of the more provocative early papers, authored by Divale and Harris, proposed that simple band and village societies are characterized by a "pervasive institutional and ideological complex of male supremacy" (1976: 521). Included as indices of this trait complex were the frequency of patriliny, patrilocality, polygyny, brideprice, male authority roles, female subordination, ideological notions of female inferiority, and male offspring preferences. The authors argued that institutionalized male supremacy is partially attributable to sexually dimorphic traits, but also to a sustained pattern of intensive warfare that placed a premium on male offspring as warrior stock. Male control over the sexual and reproductive lives of women and their distribution as rewards for bravery, they reasoned, require cultural institutions that encourage female passivity, submissiveness, and intimidation. Warfare, along with the practice of female infanticide, were viewed as indigenous population control measures.

The Divale-Harris paper in many ways represents the gold standard of male supremacist theory, in that basic sexual differences in physiology and temperament are seen as the wellspring of not only labor division, but the structure of kinship and marriage systems, the allocation of power and prestige, and of intersocietal relations. Aggression, sexual and military prowess, and endemic warfare are presented as the normative pattern for men in foraging and simple farming societies, thereby favoring male-centered social institutions and male dominion in the allocation of power and prestige. Statistical correlations between designated elements of the trait complex are presumed to represent causal relationships that are universally operative in time as well as space.

In their critique of the Divale-Harris paper, Lancaster and Lancaster (1978) questioned the universal applicability of the proposed complex of traits, such as descent, postmarital residence patterns, and marriage types, as indicators of male supremacy. The Lancasters also argued that the asymmetrical sex-linked patterns of power and prestige observed in war-like populations should not be regarded as natural extensions of sexually dimorphic traits or proclivities, but rather as responses to specific conditions such as overpopulation or resource scarcity:

> ... the male supremacist complex does not represent the entire range of human adaptive stance nor the basic pattern of relationship between the sexes. It represents behavioral responses to specific environmental conditions, conditions that have greatly increased in frequency and geographical distribution in recent millennia. The male supremacist complex is not equivalent to "human nature"; it simply describes an important behavioral response found in small-scale societies under environmental conditions in which intergroup fighting is intense. (1978: 117)

Parker and Parker (1979) saw such asymmetrical dominance patterns as arising from the unequal exchange value of male and female labor. They argued that typical female tasks were more repetitive and mundane and hence more "substitutable" in personnel, whereas male tasks such as hunting and warfare involved more individual skill and risk. Ergo, where male labor requires a higher level of achievement or carries a greater risk burden, it is more highly valued than women's work and is therefore rewarded with greater prestige and power as a kind of compensation.[6]

While acknowledging the contemporary dominance of males in the public sphere, the Parkers went on to emphasize that cultural interpretations of biological sex differences are not static, but the product of dynamic evolutionary processes driven by changing environmental conditions. In contrast to Divale and Harris, they held that technologically simple societies such as hunter-gatherers were more egalitarian due to the general absence of hostilities and the more even distribution of skills and risks in male and female labor. Accordingly, they saw the male superiority myth as a cultural ideology, one that is becoming increasingly dysfunctional in modern society as technological innovation continues to decrease the relevance of sexually dimorphic criteria for labor division.

Comments by the Lancasters and the Parkers touch on three principal areas in which innate male dominance has been challenged as a harbinger of social forms: (1) the diversity of human kinship systems and the nature and frequency of egalitarianism in foraging and hoe farming societies; (2) the bias of early ethnographic observers; and (3) an overreliance on purely synchronic or ahistorical approaches to ethnographic data.

On the first point, critics have noted that the technologically simplest human societies are not universally organized on the basis of male-centered kinship groups. Those most closely mirroring Paleolithic hunter-gatherers, at least in the ethnographic present, typically recognize broad bilateral kinship ties and maintain flexible residence patterns. Similarly, many small-scale hoe farmers are strongly matrilineal. If biology were destiny, one would expect to find universal patriliny, but this is not the case.

Second, the simplest types of human societies are also notoriously egalitarian. Egalitarianism, in this context, is not to be confused with the modern Western concept of sexual equality. Rather, it refers to the recognized *duality* of the sexes—physical, economic, reproductive, and spiritual—that is both underscored and celebrated as essential to the cycle of life. Finnegan (2017) presents an insightful summary of egalitarianism among hunter-gatherer societies, in which males and females are perceived as having distinct and reciprocal responsibilities in both the mundane and sacred domains. These disparate roles and responsibilities result in an ongoing balance of power that is regarded as necessary for the maintenance of spiritual harmony and community well-being. In a cited example, Finnegan notes that although men among the Baka pygmies are primarily tasked with providing meat, women are perceived as having a special spiritual relationship with prey animals that requires the performance of female rituals to ensure hunting success.

While sexually dimorphic traits appear to universally influence the selection of males for activities involving strength, aggressive behavior, and distant travel, labor assignments as well as the power and prestige assigned to them are highly variable. Among the aboriginal Innu (Montagnais-Naskapi) Indian foragers, for instance, both women and men were hunters and cookers of meat, pursuing game jointly as well as individually (Leacock 1954, 1955). This pattern of cooperative hunting by all community members is commonplace among foraging societies and appears to have been widely employed by hominin populations as a subsistence strategy throughout the Paleolithic. Innu women enjoyed considerable autonomy, sexual freedom, and equitable political engagement through consensus decision making. Similar patterns of female autonomy have been noted among other hunter-gatherer societies, such the Mbuti pygmies and !Kung Bushmen, who also lacked formal authority roles and instead allocated power on the basis of age and influence rather than sex (Begler 1998).

Brown's (1970) analysis of the seventeenth- and eighteenth-century Iroquois Indians provides an example of hoe farming societies in which women were the primary cultivators and enjoyed levels of personal autonomy, prestige, and power on a par with men. Sanday's (1973, 1974) cross-cultural analyses of female productive roles and social status concluded that substantial subsistence contributions were a necessary but not a sufficient cause for power allocation to women, which requires the corresponding ownership and control of strategic resources. Leacock concurred, noting that the source of this elevated status was the ability of Iroquois women, as matrons of their lineages, to "control the conditions of their work and the dispensation of the goods they produce" (1978: 232). In societies such as the Iroquois and Innu, the basic socioeconomic unit was not the patrilineage or the male-female dyad, but the uterine kin collective. Thus, there was no dichotomy or hard line drawn along sex lines between management of the domestic and public economies, which overlapped in aboriginal times. Although formal authority roles among the Iroquois were assumed by men, women exerted considerable influence in the selection of incumbents and could intervene in both internal and external political decisions through their control of the food supply. In an oft-quoted passage from *Ancient Society,* Morgan shares personal correspondence from longtime missionary Arthur Wright to underscore the high status of Iroquois women:

> Usually, the female portion ruled the house, and were doubtless clannish enough about it. The stores were held in common; but woe to the luckless husband or lover who was too shiftless to do his share of the providing. No matter how many children, or whatever goods he might have in the

house, he might at any time be ordered to pack up his blanket and budge; and after such orders it would not be healthful for him to attempt to disobey. The house would be too hot for him; and, unless saved by the intercession of some aunt or grandmother, he must retreat to his own clan; or, as was often done, go and start a new matrimonial alliance in some other. The women were the great power among the clans, as everywhere else. (Morgan 1907, orig. 1877: 455n)[7]

In summary, male dominance and the male superiority complex of traits are neither innate nor universal. Egalitarian societies illustrate both the diversity of kinship systems and the complex interplay of diffuse sources of power that may accrue to the sexes and that cross-cut the social, political, and ritual domains of a society. While the nature of kinship systems influences the nature of intersexual relationships, neither can be wholly attributed to biology. An understanding of variability in these cultural patterns requires us to move beyond dimorphic physical traits, temperaments, and reproductive roles and delve further into the ecological conditions favoring male- and female-centered kinship groups—a subject to which we shall return in the next chapter.

A second area of objection to universal male dominance theories lies in the veracity of the ethnographic data on which their conclusions are based. In the study discussed earlier, Lancaster and Lancaster (1978) cautioned against taking information on male and female status differences at face value due to potential observer bias. The majority of older ethnographic descriptions of preliterate societies (i.e., when they were least affected by acculturative influences) were conducted by male clergy, emissaries, colonial officials, and by anthropologists who arrived with their cultural values in tow. Such chroniclers were almost exclusively male, and often based their conclusions on the status of women solely on the testimony of aboriginal men. Ethnographic information has also been skewed by the a priori Western bias that the core socioeconomic unit in all societies consists, or *should* consist, of the pair-bonded nuclear family. This perspective assumes that the female sphere of action is limited to familial, domestic, or nonpublic matters, overlooking the fact that the operative unit in preliterate societies more commonly consists of a larger kin collective in which women may exert considerable influence. In a manner similar to what happened in the 1970s when women primatologists took to the field, data collected by women ethnographers and ethnohistorians has provided a more balanced and complete picture of male-female interactions and social life in egalitarian societies.

A third criticism of male dominance theories is their failure to recognize the influence of historical events, enculturation, and foreign money economies on the social structure of native peoples and, by extension, the status of women. While egalitarian relations among the sexes are commonplace among simple foragers and extensive hoe farmers, dramatic changes in female status and the nature of kinship systems have accompanied shifts in the political economy of non-Western societies in postcontact times. Many studies have chronicled the disruption of aboriginal kinship groups and the decline of female status and influence as a direct result of the penetration of foreign market economies. Leacock (1954, 1955), for example, documented how the North American fur trade intruded on and undermined the foundation of Innu kin collectives, turning male hunters into traders and introducing an external economy based on portable wealth from which women were excluded. Lancaster (1976, 1979) analyzed how intersexual relations among African matrilineal horticulturalists were gradually changed by the growth of a local cash economy, the limited productivity or loss of lineage gardens, accumulation of portable wealth, and subsequent opportunities for the purchase of jural rights in children by males. Similarly, Sacks (1975), in her comparison of four distinct types of African societies, found that the status of women declined as their involvement in collective production was eclipsed by male economic activities such as wage earning and where men were granted entry to marketing and trade positions in the external money economy.

Many other examples illustrating these historic trends appear in the literature. Such studies raise important questions about the representativeness and utility of the existing ethnographic data base for drawing conclusions about innate behavior or cultural universals. Unless information on asymmetrical sex-based dominance patterns in sample societies is placed in proper ethnohistorical context, the issue of Galton's problem resurfaces.

The Human Biogram

By the mid 1970s, sociobiological theorists began to weigh in on the question of human social origins and the nature of ancient kinship groups. As noted throughout this book, selfish-gene theories not only prioritize the extent to which biology has shaped human social forms through time, but also propose that the male genome has played the central role. In such schemes, human evolution has marched in tandem with male reproductive strategies. Because political and socioeconomic dominance by males is seen as the trademark of an imprinted human biogram, gene-based theory renders questions about flexibility in kinship reckoning essentially moot.

Such models bring specific and limited interpretations of inclusive fitness and kin selection to the analysis of human social systems. The work of van den Berghe and Barash (1977) is illustrative in this regard. They see male fitness pursuits as not only maximizing sexual access to fertile females, but parasitizing them for their reproductive potential. This is accomplished through asymmetrically binding relationships, in which males exert control over female sexuality to assure paternity and the transmission of their genes to subsequent generations. In a complementary manner, female fitness is seen as relying on the selection of mates with the greatest inventory of resources, thereby enhancing the survival potential of their offspring. Resultant human social groups are presented as inherently patrilineal on two principal grounds: (1) innate male dominance, coupled with subordinate, dependent, and submissive female roles, and (2) the greater male inclusive fitness benefits that derive from paternity assurance and investments in male offspring.

Sociobiological arguments for innate male dominance parallel those discussed above. On the selfish-gene side of the equation, van den Berghe and Barash reference hypotheses advanced by Hartung et al. (1976), who offered reasons why sons may be a better bet than daughters for ensuring a man's transmission of genes to subsequent generations. First, Hartung proposed the so-called "chromosomal corollary," namely, that the Y chromosome provides greater parental certainty to males, since it is not subject to crossing-over as is the X chromosome for females, and (barring mutation) passes unchanged from father to son. Since, by this measure, males will be more closely related genetically to their son's rather than their daughter's or sister's son's descendants, their fitness will be enhanced by transferring wealth in the male line. Second, Hartung noted that due to the nature of anisogamy, sons are biologically capable of producing more offspring than their sisters (referred to as reproductive variance)—in short, the more offspring, the more paternal genes are transmitted. Hence, it is more efficient to invest wealth in the male line.

In evaluating this model, it is useful to explore more fully Hartung's assumptions about the male fitness advantages of patrilineal descent. Notably, his chromosomal corollary holds only to the extent that just one Y chromosome is in play, i.e., if paternity at each generation can be assured (a large *if*). A shortcoming of this argument, acknowledged by van den Berghe (1979), is the doubtful significance of the Y chromosome for forging male genetic legacies since it accounts for only 1/46 of the male genome and carries comparatively little genetic material. More convincing to van den Berghe was Hartung's argument for male reproductive variance—the notion that sons have the biological ability to produce more offspring than daughters and that they can therefore provide greater reproductive returns on parental investments, particularly when coupled with polygyny.

As pointed out by commenters on Hartung's article, however, even though males are theoretically capable of producing hundreds of offspring, in practice their reproductive variance may not be greater than that of females (Mavalwala 1976; Fix 1976). That is, not all males can be assumed to have an equal opportunity to procreate, and those who reproduce may do so at the expense of other male group members. In the case of nonhuman primate polygyny, and perhaps of ancient hominid polygyny as well, the majority of males may have only limited opportunities to procreate during their lifetimes. This same exclusionary principle may also apply to humans in situations where male access to multiple mates is constrained by intrasexual rivalry or by inequality created by differential access to power and/or resources. While the passage of wealth or status from father to son may facilitate access to multiple mates, polygyny, even where preferred, may constitute a small percentage of the total unions and hence represent only a minority pathway for male fitness.

Nonetheless, having established to their satisfaction the dual prongs of a human biogram, namely, innate male dominance and patrilineal descent, van den Berghe and Barash went on to account for matrilineal kinship within the theoretical framework of male inclusive fitness (1977: 817–18). Since, they reasoned, matriliny is associated with higher frequencies of female sexual permissiveness, adultery, and divorce, this type of descent system likely originated as a male accommodation to female cuckoldry. Thus, if the probability of paternity in their wife's offspring is below 25 percent (the so-called "paternity threshold"), then males are better off investing resources in their sororal nephews rather than in their own purported sons, since they presumably share one-quarter of their genes with the children of their full sisters. In this situation, it is proposed, the male strategy for fitness maximization shifts from parasitizing their wives for the benefit of their sons to parasitizing their brothers-in-law for the benefit of their nephews. This alleged structural asymmetry in male fitness objectives, in their view, renders matrilineal societies inherently conflicted and unstable. The authors go further, noting that the price of devising such "social arrangements that buck human biology" is to perish at the hand of natural selection: "Thus the matrilineal solution, by conforming to one aspect of our biogram (male domination) created dissonance in another. As such, it was less adequate a solution than patrilineality, and it is being selected out" (van den Berghe and Barash 1977: 818).

While van den Berghe (1979: 102–8) notes in passing that matrilineal societies were probably more numerous in the past, no explanation is offered for assuming that a social form typecast as maladaptive was once more widespread, other than as an imperfect solution to female sexual promiscuity and the lack of male parental certainty (a situation that innate male dominance was apparently unable to otherwise resolve). Similarly, bilateral rather than patrilineal descent systems are proposed as the likely prototype for primeval hunter-gatherers, based on prevailing patterns among surviving foragers in the ethnographic present. But since hunter-gatherer societies account for about 90 percent of human history, at what point and under what circumstances did the proposed human biogram make itself known?

In her critique of the theoretical framework proposed by van den Berghe and Barash, Dickeman (1979) opined:

> Thus, we see here a search for "typical" forms of human social structure, which knowledge of the ethnographic record forces the authors to admit is oversimplistic. Contrary cases are then dismissed as "textbook exceptions," "reluctant adaptations," and "less adequate solutions." At points, the new sociobiology sounds remarkably like old 19th-century evolutionism: the covert moralism is the same. This is not only bad anthropology but bad biology. A more appropriate sociobiological strategy is the search, not for 19th-century classificatory types but for a specification of the range of possible behavioral responses, the ecological context that evokes each one, and the contribution of previous states (i.e., history) to current responses. This search must be phrased in terms of the reproductive consequences of each state since the fundamental premise of the approach is that human social responses are, on average, adaptive in the Darwinian sense. (1979: 351–52)

Unitary models, by definition, close the discussion on the origins and nature of human social diversity by establishing monotypic imperatives, deviations from which are interpreted as unnatural, maladaptive, and ultimately doomed to extinction. Such models also take a narrow view on the potential correlates of male and female inclusive fitness. Not only does female reproductive success take an evolutionary back seat to that of males, but the human biogram proposed by van den Berghe and Barash fails to contemplate alternative evolutionary scenarios in which the sexes may pursue their fitness in a cooperative and mutually beneficial manner. Can males successfully realize their own interests without parasitizing females? Can females achieve reproductive success in lieu of bondage to an oppressive consort? Are the reproductive strategies of the sexes inherently conflicting?

The reproductive strategies painted by van den Berghe and Barash mirror the classic pair-bonded social contract: "Men have power, and hence control over resources. Women, as do the females of many other species, choose males with the best possible resources" (1977: 815). One may well question what resources were controlled by Plio-Pleistocene males that were so coveted by females. As we have seen, there is no evidence that the evolution of human society rested on the foundation of male provisioning. The critical subsistence role of females among both nonhuman primates and human hunter-gatherers is well-established. The more likely scenario in ancient times is that females ensured the well-being and survival of their offspring not by snagging the most prolific male for economic support, but by establishing cooperative breeding relationships among their uterine kin. It was within this circle of nurturance and food sharing by mothers, sisters, and grandmothers that female inclusive fitness was maximized and that the energetics and life histories of early human groups were fundamentally transformed. Depending on local conditions, males probably made variable subsistence contributions to social group members, but survival of the offspring on which their reproductive success relied, both in terms of primary provisioning and nurturance to maturity, lay squarely in the hands of female cooperative breeders. Studies by Pavard et al. (2007a, 2007b) correlated the probability of a child's survival as a function of the mother's age at death and of the availability of secondary caregivers provided by alloparenting networks.

As noted by Smith, et al. (2010) and by Shennan (2011), parental intergenerational resource transfers to offspring prior to the end of the Ice Age consisted primarily of what Borgerhoff-Mulder et al. (2009) categorized as *embodied* and *relational* wealth, rather than *material* wealth.[8] Fitness would have been underwritten by the intergenerational transfer of embodied wealth, in the form of nutritional provisioning, socialization, and instruction in subsistence and other life skills; and of relational wealth, including membership in a social network of cooperative breeders and in alliance networks that included protective males.

Under such conditions, mate selection by females would favor cooperative over aggressive males for membership in kin collectives—males who could offer both multiple mating opportunities and group protection for the young. Hartung, in a later paper (1985: 661), in fact outlined how female inclusive fitness could be accomplished autonomously, free of the type of male parasitism outlined in some sociobiological models. Namely, in situations where females engage in multiple mating and paternal certainty is low, the probability of relatedness between a female and her matrilineal heirs is higher than her relatedness with her patrilineal heirs. Put another way, if females pursue their reproductive strategies with multiple sexual partners, then genetic relationships among uterine kinsmen will be stronger than those calculated in the paternal line(s). This opportunity for female inclusive fitness is presented by multifemale-multimale groups with permissive multiple mating patterns.

Notably, male inclusive fitness may also be advanced by these same mating and social arrangements. Male reproductive success would be optimized through increased access to multiple fertile females and through the siring of multiple offspring whose survival is underwritten by an established network of caretakers. Female-centered kinship groups with a pattern of multiple mating provide outside males all the sexual access benefits of "polygyny," plus the fitness benefits of embodied and relational wealth transfers to all their biological issue that result from unions with bonded uterine kinswomen. Multifemale-multimale communities based on cooperative rather than parasitic relationships thus allow the reproductive strategies of the sexes to become aligned, enhancing the fitness of both.

Pan-Genesis Theory

Another model that proposes to explain the origins of human kinship systems also supports the idea of phylogenetic continuity between the social life of ancient apes and modern humans, but does so with a largely synchronic analysis—one that deliberately omits references to Plio-Pleistocene ecology, inclusive fitness theory, or specific hominin lineages. In *Primeval Kinship,* Chapais (2008) draws on the kinship analyses of Claude Levi-Strauss (1967) and Robin Fox (1967) to develop a scheme of sequential structural changes in exogamic rules marking the transition from a chimpanzee-like social organization to the human family "deep structure" and unilineal descent systems. The resultant model embodies the notion of universal evolutionary stages in that its proposed developmental phases represent distinct social forms that progressed in a specific and irreversible order.

A core assumption of Chapais's ancestral male kin hypothesis is that *Pan* (and in particular *Pan troglodytes*) is a credible social avatar for our last common ancestor and the logical base from which human kinship evolved. Proto-humans are seen as mirroring contemporary chimpanzees, living in small, territorial multimale-multifemale groups characterized by promiscuous mating, female dispersal, male bonding and aggression, and intergroup hostility. Kinship recognition is largely limited to the mother-offspring dyad, and alliance systems to loosely structured bonding among localized males. With this model as a starting point, the challenge then becomes one of explaining how evolving hominins moved from biological to cultural systems of kinship—i.e., to stable breeding relationships; paternal, bilateral, and affinal kin recognition; establishment of kin group corporate identities and alliances; and the pacification of intergroup hostilities that presaged more complex levels of societal integration.

Bridging the gap between nonhuman and human kinship systems would appear to present much less of a leap with alternative primate avatars such as macaques than with chimpanzees. Macaques, for example, have well-developed and ranked matrilineages that extend mother-child kin recognition to other lineal and collateral kin over as many as three generations. But the notion that a uterine kinship legacy could have served as the progenitor of human kinship systems is discounted by Chapais on the basis that our closest living relatives, namely, *Pan*, are characterized by male philopatry. Thus, by a somewhat teleological argument, he maintains that the formation of matrilineages among ancestral humans would have been precluded by female dispersal patterns and the resultant severing of maternal kinship ties.

Chapais's alternative structural route to human unilineal descent groups (one that promotes phylogenetic continuity with contemporary patrilineal systems) instead relies on the evolution of pair bonding and biparental families. He proposes that the emergence of stable breeding relationships was initiated by males through mate guarding and body guarding, a defensive strategy that took the initial form of polygyny, followed by monogamous couplings. Pair bonding is thus seen as providing the critical linchpin that connected primeval male philopatry to paternal recognition and the evolution of agnatic kinship. The pair bond and establishment of male provisioning and caretaking roles is also credited with the sexual division of labor, female life history changes such as a reduction in birth intervals, and even setting the stage for the grandmothering of offspring.

Pair bonding is additionally proposed to have served as the precursor of bilateral kin alliances through the strengthening of brother-sister bonds and the subsequent practice of reciprocal kin group exogamy (the exchange of females). Chapais argues that while feeding competition between early human groups may have been relieved by pair bonding and male subsistence contributions, persistent male sexual competition would have required new affiliative measures for intergroup pacification. He reasons that once males were able to exercise "control" over their kinswomen, they utilized them as "appeasing bridges" or peacemakers with trading partners in neighboring groups. Female exchange is thus credited with the rise of suprakin or tribal organization.

Chapais then utilizes the premise of tribal evolution to bolster the argument that female philopatry and matrilineal descent groups are both latecomer developments and outlier phenomena in human populations. He reasons that social organization based on male dispersal is not only incompatible with the *Pan* biogram, but could not have emerged prior to the achievement of intergroup pacification, i.e., when males could travel across local territorial boundaries unmolested, free to maintain maternal kinship bonds with their sisters and nephews.

However, the *Pan*-genesis theory, as a Paleolithic legend, quickly unravels when one modifies its fundamental premise or starting point. The notion that our last common ancestor and all of its hominin descendants were cut from a chimpanzee-like cloth contradicts much of what we know about the impact of ecology on primate social organization, as well as the disparate evolutionary trajectories of extant species. Philopatric and behavioral patterns are known to reflect the type and distribution of resources and female feeding strategies in specific types of niches. For example, the dispersed distribution of small, high-quality food patches in the homelands of *Pan troglodytes* engenders intense competition and militates against female bonding and joint social feeding. In contrast, critical food resources such as leaves and large fruits are more plentiful and evenly distributed in the habitat of bonobos or *Pan paniscus,* who exhibit strikingly different social behaviors—namely, female dominance, female bonding and social feeding, lifelong mother-son alliances, the absence of strong male intrasexual bonding, and larger social groups (de Waal 2005: 65–72; 2013). As noted in the previous chapter, the fact that behavioral differences between these closely related *Pan* lineages is marked by measurable differences in the neurobiological structure of their brains suggests that they have had distinct adaptive histories.

The forested refugia currently occupied by both chimpanzees and bonobos are, in turn, distinct from the more open, mosaic Plio-Pleistocene landscapes in which *Homo* is now thought to have evolved, settings more likely to contain the type of dense, high-quality food patches that favor more intensive patterns of female social feeding. Among non-forest-dwelling primate species occupying habitats with this type and distribution of food patches, female philopatry and permanent, multigenerational, ranked matrilineages predominate.

A reasonable argument can be made, therefore, that the earliest humans may not have been chimp-like at all, but instead may have lived in large, nonterritorial multimale-multifemale communities composed of highly structured kinship groups that clustered related females. It has been suggested that this type of ecological and social setting would have selected for the type of mosaic brain evolution and capacity for behavioral plasticity that enabled early humans to successfully break through the "gray ceiling" and forever separate themselves from their distant primate cousins.

Let us assume, for the sake of argument, that our forebears began from this alternative biobehavioral platform rather than that of contemporary chimpanzees. Could distinctively human kinship systems have evolved from this base? Chapais says no on the basis of three platform assumptions: (1) male philopatry is genetically imprinted, (2) pair bonding and paternal recognition *must* precede the development of unilineal kinship groups, and (3) tribal organization *must* precede residential diversity and the evolution of matrilineal descent.

As just noted, primate philopatry appears to be primarily dictated by female feeding strategies, which are in turn dictated by the nature and distribution of resources. Contemporary nonhuman primates tend to have fairly rigid mating and dispersal patterns whose genetic underpinnings have been shaped by adaptations to specific niches. Human patterns, while founded on the same selective principles, have been shaped by not one, but a variety of niches. The greater plasticity and ecological range of ancient humans was enabled by mosaic brain reorganization, the emergence of flexible reproductive strategies, and the imprinting of epialleles that calibrated adaptive responses to stochastic conditions. An alternative hypothesis, then, is that unlike modern chimpanzees, early hominins were probably not locked in to a single philopatric mode, but rather were equipped with a broader menu of social technologies.

With regard to the structural antecedents of unilineal descent groups, Chapais concedes both the prevalence of uterine kinship links among contemporary nonhuman primates and the fact that such alliances frequently provide the foundation for multigenerational matrilineages. He is careful, however, to characterize these kinship units as simply uterine affiliations or kindreds rather than true descent groups. Nonhuman primate matrilineages, though commonplace, are summarily dismissed as "embryonic" or "metaphorical" descent groups due to the absence of paternal recognition and biparental pair bonds.

For Chapais, as for many other hominin social reconstructionists, it is difficult to envision the existence of a viable human kinship group without the pivotal involvement of a designated and exclusive male progenitor. The idea of "stable breeding relationships" is conceptually equated with the pair bond, consisting of one male and one or more females. In the absence of established male family heads, it is argued, offspring would be unable to recognize their fathers or the concept of bilateral affiliation. However, studies of kin recognition among primates with matrilineal social groupings, such as rhesus macaques and baboons (*Papio cynocephalus*), suggest that individuals not only recognize paternal kin, but exhibit discriminatory behaviors toward maternal relatives, paternal relatives, and nonkin (Alberts 1999; Widdig et al. 2001; Smith, Alberts, and Altmann 2003). While such biases in affiliative behavior may be influenced by age proximity and recognition of parental phenotypic traits, there is evidence to suggest that genomic mechanisms, namely, imprinted genes on the maternal and paternal X chromosomes, play a contributing role in kin recognition (Isles, Davies, and Wilkinson 2006: 2233–34).

The alternative argument presented here is that early human families were likely not biparental, but *multi*parental. Early hominin communities were composed not of aggregate clusters of male-female dyads, but rather of socioeconomic kinship units that coalesced as an extension of female collaborative breeding. This social context provided the intensive and sustained interaction among related females and their offspring necessary for the recognition of both lineal and collateral kin and for the strengthening of sibling relationships. The dramatic changes in hominin life histories and the unique reorganization of hominin brains relied on the combined support network of multiple mothers, grandmothers, and partible fathers. Once again, it took a village.

Chapais makes the assumption that ancient female philopatry would have been incompatible with the genesis of true unilineal descent groups since, under the presumed conditions of intergroup hostility, the dispersal of males at adolescence would have the effect of severing their relationships with natal uterine kin. Hallmarks of human matriliny, therefore, such as durable mother-son/brother-sister bonds and avuncular alliances, could not evolve. Notably, however, nonhuman primate societies organized around uterine kin do not conform to the chimpanzee model of small, competitive, atomistic bands. Instead they aggregate into larger nonterritorial communities containing multiple matrilineages. Arguably, this type of local group structure provides the basic architecture for what Chapais calls suprakin or tribal organization, in which component outbreeding kinship units have established stable, noncombative relationships (in this case, the shared and allocated exploitation of a given niche). Under these conditions, male uterine alliances could be maintained through the reciprocal exchange of dispersing kinsmen between the component lineages, i.e., through the combination of local endogamy and lineage exogamy. This, in fact, mirrors the pattern of human matrilocal/uxorilocal societies, which typically organize closely spaced matrilineages into composite multikin communities, on either a seasonal or permanent basis.

Thus, if primeval kinship models begin with the alternative platform of female philopatry, the sequence of evolutionary phases proposed by Chapais is fundamentally changed, if not reversed. Namely, stable mating relationships are first established in the form of partible paternity within the component cooperative breeding groups of a community. Such groups give rise to durable lineal and collateral kinship bonds and coalesce into multiple matrilineages. In this scenario, pair bonding is the product rather than progenitor of reciprocal exogamy. The male-female reproductive dyad serves as a mechanism for joining *not* individuals, but kinship groups into complex bilateral and affinal alliance networks. Recognition of paternity becomes an incidental but unnecessary element in this process, and only rises to significance if subsequently utilized as a criterion for kin group affiliation and the transfer of intergenerational wealth. As Chapais proposes, the formation of patrilineal kinship groups is based on a reinvention of uterine affiliation principles and is preceded by male philopatry and the localization of agnatic kin. Whereas Chapais sees patriliny and male philopatry as evidence of our phylogenetic continuity with *Pan*, this alternative model credits epigenesis with the capacity of evolving hominins to organize their communities in multiple ways, calibrating their kinship strategies with ecological conditions.

Chapais footnotes his primeval kinship model with the statement that nineteenth-century evolutionists such as Lewis Henry Morgan (1870, 1907, orig. 1877) and John McLennan (1865), who advocated for ancient matriliny, would have discovered the error of their ways had they only known about our *Pan* cousins and their penchant for male philopatry: "They, of course, had no reason to assume that before humans formed pair bonds, males were localized while females dispersed, a situation that precluded matrilineal descent. Had they known this, they would have had to conclude that patrilineality came first" (Chapais 2008: 297). But for other modern evolutionists, innate male philopatry is neither an obvious nor foregone conclusion.

Participants in the 2005 Royal Anthropological Institute, London workshop on "Early Human Kinship," for example, revisited early evolutionary models on ancient social organization in light of contemporary ethological and ethnographic information. Several papers delivered at that workshop and published in a collection under the same title (Allen et al. 2008) witnessed not only the renewed dialogue of British social anthropologists on human evolution, but a reopening of the debate on primeval kinship. In the introduction to this collection, James notes:

> It is probably helpful to concentrate on the likelihood that physical motherhood would be the only point of individual reference in the earliest systems; and that "fathers" would be best understood under sociocentric categories. All the evidence one could assemble for the likely circumstances of a "starting point" for human kinship indicates that we should not expect to find symmetry in relations between male and female, nor genealogical specification of individual fatherhood. (2008: 15)

While some workshop contributors, such as Gowlett (2008), continue to emphasize the importance of pair bonding for the emergence of *H. erectus*, others question the notion that paternal certainty, pair bonding, male provisioning, and male philopatry were necessary precursors of early human social organization. In contrast, participants such as Knight (Knight and Power 2005; Knight 2008) and Opie and Power (2008) join American researchers such as O'Connell et al. (1999) in arguing that the platform of female dispersal and female subsistence dependency on consorts is incompatible with the sea change in female reproductive energetics that would have been required to propel encephalization and the evolution of *Homo* life history patterns. Instead, they advocate strongly for the central role played by female kin coalitions for coprovisioning and alloparenting. Based on their modeling of *H. erectus* female energetics, Opie and Power conclude that ancestral hominids, from *H. erectus* to early moderns, were naturally biased toward matrilocality and matrilineal kinship:

> The results suggest that female *H. erectus* could not rely on her mother's help alone, or on the help of a pair-bonded male alone, if she were to raise enough offspring to replace the population. Females would have needed to draw on both these sources of allocare and help with provisioning. This implies that a *H. erectus* mother would need to live throughout most of her reproductive lifespan with female kin. Any tendency to male philopatry in early *Homo* would cut females off from such kin support. By contrast, a female philopatric model allows "man/hunter/scavenger" to play his role. Grandmother and man the cooperative scavenger become mutually reinforcing, able between them to provide children with regular supplies of energy and high-quality nutrients. (2008: 186)

In summary, theories on the nature of primeval kinship make fundamentally different assumptions about potential phylogenetic continuities in the social patterns of humans and contemporary apes. What have been referred to here as *Pan*-genesis models see male philopatric patterns in chimpanzees as evidence of genetic imprinting that emerged millions of years ago in our last common ancestor and that presaged the evolution of human patrilineal descent systems. Alternative models of ancestral kinship lay no such claim to a primal social biogram. Instead, they rely on an understanding of how female feeding and reproductive strategies influence the structure of primate social groups, now and in the ancient past. Such models recognize that hominin sociality patterns are plastic and that they were importantly shaped by female energy expenditures required for both the provisioning and nurturance of offspring in a given niche. Models that support a bias toward ancient female philopatry base this conclusion on the adaptive advantages gained by cooperative breeding for meeting the extraordinary energetic requirements of *Homo* encephalization and life history changes.

If female kin collectives provided a social structure that enhanced the reproductive success of ancient hominins, how did natural selection operate to favor this pattern? New genomic research has provided another perspective on the inclusive fitness impacts of sex-biased dispersal and the disparate roles played by the sexes in the trajectory of human evolution.

Genomic Imprinting, Kinship, and Inclusive Fitness

Theories on the origins of human kinship systems have relied rather heavily on armchair anthropology, fashioning biological rationales for human social forms based on what are perceived to be ancient primate templates. The fact that modern humans share about 98 percent of their genome with chimpanzees has reinforced the popular notion that the social life of these contemporary forest-dwelling apes mirrors that of our last common ancestor. With so much shared DNA, what difference could 2 percent really make? The answer appears to be: *night and day*. The *Pan-Homo* split some 5 million years ago was marked by genomic mutations that coded for dramatic changes in brain evolution, life histories, energetics, and sociality. Groundbreaking research in evolutionary genetics and neurobiology indicates that the male and female genomes played disparate roles in this process and that hominin evolution both relied on and was accelerated by the formation of social groups based on common kinship.

Insights into the selective processes by which these developments occurred in the evolving genus *Homo* are being provided by recent studies of maternally and paternally inherited *imprinted genes* (epigenetically marked genes in which only one allele is expressed and one allele silenced, by parent-of-origin). Imprinted genes, which are closely associated with the control of pre- and postnatal growth, brain function, and social behavior, account for about 150 of the estimated 19,000 to 20,000 DNA coding genes in the human genome. Studies have focused on the variable contributions made by these maternal and paternal imprinted genes to functionally distinct regions of the primate brain and the types of social contexts that may have accelerated the differential development of these brain regions through time.[9]

Two brain regions are of critical importance in hominin encephalization: (1) the frontal cortex (neocortex and striatum) or "executive" brain, which is associated with cognitive functions, and (2) the limbic (hypothalamus and septum) or "emotional" brain, which is strongly influenced by hormones and associated with "primary motivated behavior" such as sexuality and parenting. Mammalian genomic studies indicate that development of these brain regions is guided by the expression of imprinted genes. Specifically, the maternal genome is linked to development of executive brain regions, while the paternal genome is represented in regions of the emotional brain, such as the hypothalamus. Associated evolutionary trends in brain evolution have important social implications. Keverne and colleagues observe:

> While certain regions of the primate forebrain (neocortex, striatum) have expanded relative to the rest of the brain, other forebrain regions have contracted in size (hypothalamus, septum). Areas of relative expansion are those to which the maternal genome make a substantial developmental contribution. This may be significant with respect to the importance of primate forebrain expansion in the development of complex behavioural strategies and the way in which these are deployed, especially in the matriline. In many primate societies the maintenance of social cohesion and group continuity over successive generations is dependent on the matriline, with high-ranking females producing high-ranking daughters that stay within the group. Regions of relative contraction are those in which the paternal genome makes a differential contribution and these are target areas for gonadal hormones, which is congruent with the diminished role for gonadal hormones in the emancipation of primate reproductive behavior. (Keverne et al. 1996a: 689)

Selective pressures on constituent parts of the brain led to the gradual freeing of reproductive behaviors from the strict dictates of gonadal hormones, allowing both sexes to benefit from the broader range of social strategies provided by increased cortical control. Significant outcomes for reproductive activity included parallel sexual receptivity for males and females, the attenuation and displacement of male aggressive behaviors with alternative social alliances, and the extension of maternal nurturant behaviors to alloparenting among close female kin.

Davies, Isles, and Wilkinson (2005), along with Haig (1997), propose that maternal and paternal genomes have conflicting interests in the allocation of resources to their offspring. The juxtaposition of male and female reproductive strategies is envisioned as playing out at the genomic level, where the conflicting interests of maternally and paternally expressed genes are cast in antagonistic roles. The "genetic conflict" or "kinship" theory holds that paternally expressed genes promote placental, embryonic, and early postnatal growth in offspring by maximizing the extraction of resources from the mother. A father's reproductive success is thereby enhanced in relation to that of other males who may subsequently impregnate the same female. In contrast, it is theorized that the interest of the maternal genome is to maintain a reproductive balance by resisting the exhaustion of a mother's resources on a single pregnancy. By allocating resources equally to future offspring, the lifetime fitness of maternally derived genes is maximized.[10] In multimale–multifemale social groups, where the offspring of females are sired by multiple males (either within the brood or across a female's lifetime), inter-genomic conflict between maternal and paternal imprinted genes is seen as an extension of their effects in-utero. Namely, paternal genomic interests are maximized not only by the flow of nutrients to the fetus across the placenta, but by the signaling of post-partum maternal behaviors for suckling and the extended nurturance of offspring. Conversely, the mother's fitness is advanced by sharing available resources equally among all of her current and future issue, who will carry forward the legacy of her maternally-derived genome.

Other theorists have discounted the notion of parental genomic conflict. They observe that although maternal and paternal genomes play complementary, interactive roles, the regulation of imprinted gene expression is primarily under matrilineal control, and has evolved primarily to calibrate both in-utero and postpartum development of offspring with resource availability (Keverne et al. 1996a, 1996b; Keverne and Curley 2008). Keverne (2017) notes:

> Perhaps a better way of conceptualizing these particular imprinted gene actions is to consider such early imprinted genetic events, which occur during in-utero development, as being co-adaptive. They serve as a "rheostat" to optimise a balanced foetal growth that is in the interests of both mother and offspring. Moreover, imprinting and parental conflict does not make sense when the control over imprinting for these genes is maternal, and any impairment to in-utero growth, or to placental development is a high risk for survival of both mother and offspring. (66)

The haploid dominance of maternal imprinted genes is critical to hominin encephalization in two ways: (1) the postponement of significant brain growth in offspring to the post-partum period, and (2) the requirement for extended care in which maternal behaviors play an enduring role in offspring brain development. According to Keverne and Curley:

> For the developing infant the mother provides the most significant environmental influence, shaping offspring brain development by producing long-term epigenetic modifications to neural and behavioural phenotypes. Mothers behaviour is able to influence the development of brain regions that are important to future regulation of maternal care in their daughters and boldness in their sons. Thus non-heritable epigenetic modifications enable long-term stable changes in neural and behavioural phenotypes in response to environmental experiences. Conceptually these experiences have much in common with learning and memory, but differ in the timeframe whereby early life experiences may impact upon the behavioural phenotype at later periods in life. (2008: 409)

If mosaic brain evolution is importantly linked to maternal imprinted genes, what role did primeval hominin kinship play in this process? Imprinted genes are thought to evolve where asymmetries of relatedness exist among the members of a social group, i.e., where there is a prevailing pattern of sex-biased dispersal from natal social groups at puberty. Philopatry creates asymmetries of relatedness, ensuring that group members share not only a large percentage of alleles with their nondispersing consanguines, but a variety of social behaviors as well. In the example of extended nurturance noted above, the fitness benefit to maternally derived alleles associated with brain development is enhanced where female kin remain together for alloparenting, spacing their individual reproductive efforts to facilitate the shared care of offspring (Isles et al. 2006).

Haig proposed that inclusive fitness theory needed to be reassessed in light of imprinted gene effects, since they produce haploid dominance rather than the typical Mendelian expression of autosomal alleles (Haig 2000; Wilkins and Haig 2003). When combined with sex-biased dispersal patterns, imprinted genes have potentially distinct consequences for the direction of evolutionary change and therefore have provided a new window on the nature of early human kinship groups:

> Autosomal alleles of either parental origin, carried by either sex, have a 50 percent chance of being transmitted to the dispersing sex in the next generation, whether this sex is male or female. Therefore, patterns of autosomal relatedness will be broadly similar in social groups based on male and female philopatry, because fathers have the same degree of relatedness to sons as mothers have to daughters. Despite this symmetry, average degrees of relatedness would tend to be higher when females remain with their natal group and males disperse because the females

will usually have greater confidence that they are mother and daughter than two males will have that they are father and son. Uncertainty of paternity therefore becomes one factor that may favor social systems based on matrilineal descent. (Haig 2000: 157)

While female X-linked alleles have an equal probability of being transmitted to offspring of both sexes, the X-linked alleles of males are *always* transmitted to daughters. This, Haig proposes, creates greater degrees of relatedness within and between generations. In short, if mosaic brain evolution, and corollary changes in behavioral plasticity, life histories, and cognition relied in large part on the haploid dominance of the maternal genome, then hominin primeval kinship would have favored female philopatric groups.

An additional factor in mosaic brain evolution that highlights the potential role of the maternal genome in evolving *Homo* is natural selection for high rates of energy production. Dramatic changes in human energetics were marked not only by the up-regulation of energy production genes in the neocortex (Marino 2006), but to a complex of parallel life history changes in evolving hominins. Mitochondria (mtDNA), the powerhouses of human cells, number by the scores in the body's eukaryotic cells (those with a visible nucleus), and have their own genome separate from the genome of the cell nucleus. Mitochondrial genes do not reshuffle during the sexual production of gametes, but rather are transmitted through maternal inheritance. This is accomplished during embryogenesis by an enzyme (a "self-destruct button") that degrades sperm mitochondria after fertilization (van der Blick 2016). Pennisi (2016) has recently suggested that mitochondria may indirectly play a major role in evolutionary change due to their high replication and mutation rates.[11] This would suggest that, as among maternal imprinted genes, selection for the expression of variable mitochondrial alleles could be similarly impacted by sex-biased dispersal patterns.

In his recent book, *Beyond Sex Differences: Genes, Brains, and Matrilineal Evolution,* Keverne takes to task prominent nineteenth-century scholars such as Huxley, Darwin, and Broca for promulgating false narratives about the nature of the sexes and their respective roles in human evolutionary success. He concludes:

> At whatever level we consider female biology, be it starting at our genetic germ cell origins and finishing with the immense complexity of the human brain, we find it is the female genome/epigenome that has been integral to, and foremost in, ensuring the success of this evolutionary progression. Specifically, it is the matriline that has taken a lead for the many genetic and epigenetic developments that have given rise to the successful evolution of the sex differences as we recognise them today . . .

Of course, the autosomal maternal and paternal genomes have a matching and necessary role in supporting the success of these developments, but from an evolutionary perspective, the selection pressures for change have operated primarily through the matriline. (2017: 173)

Genomic studies thus underscore the central role played by the matrilineal genome in hominin brain evolution and human biological success, and lend support to the theory of female philopatry in ancestral social groups. In the spirit of Chapais's indictment of Morgan and McLennan cited earlier in this chapter, if the architects of contemporary androcentric theories had been informed of and guided by recent findings in the fields of genetics and neurobiology, they may have been compelled to draw quite different conclusions. Uterine kinship groups constitute a stable, affiliative intergenerational structure in which complex decision making, cooperative social behaviors, and aggressive constraints could have evolved (Smuts 1987; Keverne 1992). The aggregation of these kinship units into multimale-multifemale breeding communities provides a model socioeconomic context in which selective pressures for resolution of disparate male and female fitness strategies were exerted—a resolution that ultimately resulted in a fundamental remodeling of the human brain and the cognitive architecture for corporate group life.

Conclusion

What have been described in this chapter as Paleolithic legends are ideas that have biased or narrowed our perception of ancient social life. Whether these theories are based on discrete evolutionary stages, innate dominance patterns, human biograms, or phylogenetic continuity with chimpanzees, they have compromised our understanding of human adaptations through time. If the pathways to inclusive fitness and the origins of human kinship systems are to be approached more broadly, monotypic models must be replaced by theoretical frameworks that recognize behavioral plasticity, epigenetic phenotypes, and the variable responses of human populations to dynamic environmental conditions. Genomic studies suggest that patterns of sex-biased dispersal and the nature of hominin kinship groups have had an ancient and enduring nexus with energy capture and the intergenerational allocation of resources. The concluding chapter of this book is devoted to the exploration of kinship as a social technology for hominin niche construction and to alternative theories on the antecedents of matrilineal, patrilineal, and bilateral social systems.

 7

Kinship as Social Technology

Recognition of common kinship as an organizing principle for cooperative socioeconomic groups is perhaps the most ancient, universal, and persistent hominin trait. The primate pattern of sex-biased dispersal from one's natal group at maturity results in the localization of kinsmen related in either the uterine (female philopatric) or agnatic (male philopatric) line. For most of human evolutionary history, such unilateral kindreds or unilineal descent groups composed the core social networks through which individuals interfaced with one another and with the material and nonmaterial worlds.

This chapter approaches Paleolithic kinship systems as social technologies—that is, as dynamic organizational strategies that fostered successful adaptations in response to environmental externalities. For millennia, humans had only a simple set of material technologies at their disposal for procuring food, shelter, and the other necessities of life. The primary survival tools of ancient peoples were thus not material, but social. Strategic kin-based social networks were relied on to structure access to resources, organize labor and food distribution, nurture the young, transmit essential knowledge and information, and regulate reproduction. Male and female philopatry provided alternative organizational templates for such networks and had disparate consequences for social demography, mating patterns, political economies, and intergroup relations. Both of these organizational options were available to and variously utilized by Paleolithic peoples in their efforts to adapt to changing environmental challenges and opportunities. It is this very plasticity that defines our hominin biogram and that fostered the course of gene-culture evolution.

This chapter will attempt to unravel the tangled web and long-standing debate among cultural anthropologists and sociobiologists on the antecedents of matrilineal, patrilineal, and bilateral kinship systems.[1]

The Essence of Kinship

In perusing recent programs for the American Anthropological Association annual meetings, it is rare to find symposia on kinship theory or human social origins. These are subjects on which previous generations of anthropologists spent much time and energy researching, debating and, occasionally, engaging in venomous exchanges. Kinship studies have been one of the hallmarks of the discipline, whether couched as archetypes in early evolutionary schema, sliced and diced in a long line of ethnographies, or reduced to the selective altruism of selfish genes. Sustained interest in kinship theory today is therefore conspicuous by its absence. It is almost as if a truce has been called on continuing intellectual discourse—that advocates of competing theories have separately decided that all the important questions have already been answered or, alternatively, that they have simply agreed to disagree. Jones (2000) observes:

> I believe that both cultural anthropologists and sociobiologists have neglected the study of kinship in part because it conflicts with their respective origin myths. Cultural anthropology's origin myth is the story of how a plucky band of cultural anthropologists slew the dragon of reactionary biological determinism in the early 20th century... For sociobiologists the origin myth is the story of how a plucky band of individual selectionists slew the dragon of sloppy good-of-the-species thinking in the 1960s and 1970s. (2000: 805)

Jones goes on to note that advocates on both sides of the human kinship equation have lodged valid indictments of the other, but in the process have often thrown the proverbial baby out with the bath water. For the kin selection purist, altruism is simply a biologically driven means to an end—a gene's passport for replicating itself over time. In this scenario, the nature of human kinship is essentially predetermined by factors seen to enhance the efficiency and success of selfish genes on their reproductive journey. The flip side of this concept is that the process of human evolution is more of a two-way street. Namely, sociality may in turn drive genomic change through a dynamic feedback loop between genetic propensities, behavioral plasticity, and the epigenetic transmission of information to succeeding generations. As noted in previous chapters, support for this interactive theory is found in recent studies on the mosaic evolution and neurological structure of the human brain. Given the long history of debate, it is perhaps time to modulate the philosophical pendulum swings and begin to construct an integrative theory of human kinship based on the foundation of gene-culture co-evolution.

Jones (2000) argues that the concept of individual kin selection needs to be expanded to include socially imposed acts of unreciprocated altruism among mutual kin, or what he refers to as "group nepotism." In other words, humans pursue their fitness within the social context of kin solidarity groups that function above the individual level, acting collectively for the mutual benefit of their members. Such collective action forms the basis for emergent moral sentiments based on notions of the public good. Jones suggests that kin groups that develop and enforce "group-beneficial values" may be expected to enjoy greater reproductive success than those that rely solely on individual altruism or selfish acts, the benefits of which fade with distance from the immediate family unit.

Gintis (2011) also takes a more integrative gene-culture approach. He notes that whereas the genome encodes environmental factors that are relatively stable over time, the human capacity for phenotypic plasticity provides a mechanism for the epigenetic (i.e., learned or cultural) transmission of fitness information among group members and across generations. He sees altruism as an emergent property of gene-culture co-evolution, noting that humans have a genetic propensity for other-regarding behaviors, such as cooperation and the internalization of moral values and norms. Such integrative models are compatible with the concept of multilevel selection in that kin-based social networks are viewed as integral to individual fitness not only by solidifying relationships with immediate blood relatives, but by structuring the allocation of corporate benefits that accrue from membership in the larger collective.

What, then, is the essence of kinship? In the broadest sense, kinship is a system of social rules that serves to optimize reproductive success for the members of a breeding population in a given niche. These rules have multiple and interactive foundations, including genes, epialleles, and culture. Their phenotypic expression is reflected in social networks established for the acquisition, processing, and allocation of fitness-related resources to group members. The structure of such networks is dynamic and responsive to stochastic environmental conditions.

In contemplating societal origins, Allen (2008) created a theoretical model of the simplest possible type of human kinship system based on elementary rules for both mating ("horizontal relations") and recruitment ("vertical relations"). Exercise of these rules by ancient hominins would create a dual division within society approximating a moiety system. Within these classificatory divisions, Allen proposes that further descriptive distinctions would be required to maintain the prohibition on incest, namely, those eliminating one or another category of primary relatives. These additional rules would thus result in the creation of four sections, or "tetradic" sociocentric divisions, as the minimum structure needed to regulate mating and the intergenerational perpetuation of social groups. Such a simple kinship system, he argues, could have originated in early hominin fission-fusion groups, where cross-cutting generational and recruitment moieties may have been ritualized during phases of composite-group aggregation. He proposes further that the evolution of simple human kinship systems may have been possible in the absence of language, since section membership could have been marked by other, nonlinguistic means (i.e., body painting, ornamentation, or ritual behaviors). Allen (2008: 111) notes: "After all, to operate the system in a minimal manner, ego only needs to know two things: into what section she can marry (horizontal dimension), and in what section her children belong (vertical dimension)." Although social networks in the current ethnographic record are more complex, Allen suggests that these elementary rules and this type of baseline structure is inherent to all human kinship systems.

Hamilton and colleagues (2007a, 2007b) focused on the phenotypic expression of kinship rules by studying the structure and fractal-like scaling of social networks in hunter-gatherer societies. Their work brings a perspective on human kinship that is compatible with general systems theory. Namely, complex hunter-gatherer social networks are seen as structures that serve to acquire, distribute, and transform "fitness-related resources," such as energy, materials, genes, and information, within a society or system. Kin groups function to maximize fitness by optimizing the flow of these commodities to their members. Individuals form hierarchical or nested social groups that arrange themselves spatially and may change in size and distribution in relation to environmental conditions. Hamilton et al. note:

> Cohesive forces in hunter-gatherer groups include kin selection due to genetic relatedness, sharing of non-genetic information and exchange of material resources. There are clear cohesive forces within families and wider kin relations, but there are also cohesive forces that extend to larger groups at higher levels of the societal hierarchy. These include exchange of marriage partners so as to avoid interbreeding, communication of information about social and environmental conditions, and exchange of material resources through trade and commerce. (2007b: 2200)

Hominins are not solo or independent breeders, but group-bonded mammals that rely on participation in cooperative social networks for their survival and reproduction. Such multilayered networks, interwoven by the idiom of kinship, were the dominant form of social organization during the formative millennia of our species' development. Male and female philopatry are common to many primate species and hence represent very ancient patterned responses to ecological conditions. What is unique to hominins is the elaboration of these alternative organizational principles to create complex, cooperative socioeconomic and reproductive groups with greater permanence in both space and time.

So adaptive and resilient have uterine and agnatic kinship groups been as core social technologies that they have survived, at least in principle, from the Plio-Pleistocene through the historic period.

The reasons why Paleolithic peoples chose to organize their social life around male- or female-related kin—why either the paternal or maternal line was prioritized to the exclusion of the other—has been the subject of scientific debate for over 150 years. The fact that this discussion has waned in recent years perhaps indicates that scholars are either flummoxed by the elusiveness of predictive explanations or, as Jones has suggested, they have simply become vested in their own origin myths. More than a half century ago, Goodenough (1963), in his review of the classic work *Matrilineal Kinship* (Schneider and Gough 1961), expressed reservations about the extent to which various factors identified by the authors as correlative with matriliny were in fact predictive:

> I do not think that the variables selected by Gough and Aberle are in themselves the immediate determinants. They are relevant, but at one or two times removed. In order to develop a more satisfying theory of descent groups, I think we must turn for the time being from our traditional classifications of modes of subsistence, levels of productivity and political centralization, and the like, and consider the kinds of problems, as defined by human purposes and human interests, that descent groups help to solve. (1963: 925)

Goodenough went on to suggest that primary attention be paid to the contrastive problem-solving strategies pursued by matrilineal and patrilineal societies in areas such as intergroup relations, allocation of the means of production, and the development of a productive workforce. He proposed, in essence, that a more fruitful approach to understanding unilineal descent systems is to set aside, at least momentarily, our favorite origin myths and instead look at the ways that uterine and agnatic organizations structure the fulfillment of basic human needs.

Conditions Favoring Matrilineal Organization

The notion that kinship groups formed on the principle of uterine linkage may constitute an evolutionarily stable strategy—one capable of supporting mutually compatible and complementary male and female reproductive goals—has received comparatively little attention in sociobiological theory. Evolutionary reconstructions have generally treated matriliny as *non*patriliny; that is, as either a primitive social form or as a deviant, specialized departure that must be explained as an exception to the rule. This perception has been underscored in sociobiology by a theoretical skew toward male fitness. As noted earlier, this bias has led to what may be dubbed the "loose women" or "when-the-cat's-away" theory of matrilineal origins, which proposes that uterine kinship groups emerge as a male accommodation to the lascivious and untrustworthy sexual nature of females. Such theories ignore the obvious fact that female multiple mating patterns, by definition, have male beneficiaries, and that matriliny may structure such behaviors in a way that enhances the fitness of both sexes.

The perception of uterine kinship as an aberrant form of social organization has also been a recurrent theme in American ethnology, which has traditionally declared male-centered kinship groups as the human default and matriliny as the curious outlier. Matrilineal origin theories based on this premise seek to identify independent variables that account for what are regarded as essentially unnatural, and hence unstable, social institutions and relationships. The comparative rarity of matrilineal societies in the ethnographic present has frequently been submitted as evidence for this claim. For example, cross-cultural studies (Aberle 1961: 663) have pegged the frequency of matriliny at 15 percent, while patriliny accounts for 44 percent of societies coded in the World Ethnographic Sample. But as pointed out many years ago by Dickeman (1979), contemporary frequencies should not be interpreted to mean that matrilineal social organization is a quirk, or maladaptive in an evolutionary sense. Rather, matrilineal rarity must be understood as a function of *niche* rarity.

The approach taken in this book parallels that of Hamilton et al. (2007b: 2196), namely, that human social organization is a tool for niche construction that has evolved "to optimize the acquisition and distribution of fitness-related resources to group members." Accordingly, selective factors favoring matrilineal and patrilineal kinship systems are manifested in the way they structure reproductive and socioeconomic relations, and the conditions under which each strategy is adaptive. The assumption here is that social strategies, like the environmental settings in which they are pursued, are dynamic, and that natural selection, like the fabled blind administrator of justice, is an agnostic process that plays no a priori favorites.

This section of the chapter will review theories on the ecological correlates of matrilineal systems, as well as the destabilizing factors that lead to structural asymmetries and the erosion of uterine kinship groups. It will forward the view that matriliny represents an adaptive, stand-alone social technology for niche construction that was more widely represented in the ethnographic past than it is today.

Subsistence and Labor Contributions

Several unsuccessful attempts have been made over the years to link patterns of residence and descent to the division of labor by sex, i.e., to establish that uterine and agnatic organization are determined by the relative productive dominance of women and men, respectively.[2] Others have proposed that the pivotal factor is not subsistence dominance per se, but the extent to which the labor of the sexes is collaborative. Gough (1961a: 553), for example, concluded that matrilocality is favored where women cooperate in small teams and where production sites are either stable or dictate group movements. Service (1962: 67, 121) also regarded collaborative labor as key, and, on that basis, proposed that matriliny was limited to horticultural societies, being entirely absent among hunter-gatherer bands.

Other studies have discredited female subsistence dominance and collaborative labor as the precursors of matrilineal systems, along with the limitation of uterine social organization to horticultural societies. Aberle's (1961) cross-cultural study documented the historic concentration of matrilineal societies among hoe farmers, but also noted their presence in "extractive" (hunter-fisher-gatherer) subsistence types. My own earlier research also confirmed the broad representation of matrilineal organization among foraging societies, as well as the existence of an inverse relationship between uterine organization and the subsistence contribution of males (Martin 1969, 1974). Similar findings for hunter-gatherers have been reported more recently by Marlowe (2004). Finally, studies by Ember and Ember (1971) and by Divale (1974) found no statistical confirmation for the female subsistence contribution hypothesis. These results, along with the fact that women often play a dominant role in the subsistence of both matrilineal and patrilineal hoe farmers, have led investigators to look for other variables that may influence the structuring of kinship groups.

Warfare, Migration, and Matrilocality

A series of cross-cultural studies conducted in the 1960s and 1970s established connections between postmarital residence patterns and the prevailing type and frequency of warfare. Van Velzen and van Wetering (1960), Otterbein and Otterbein (1965), and Ember and Ember (1971) all linked feuding and internal warfare with the presence of "fraternal interest groups," or the local aggregation of related males. In contrast, societies with matrilocal residence, which disperse male agnates by rule, were found to have internally harmonious relationships and to limit their military efforts to warfare with external sociopolitical groups.

Thus, the way in which local kinship groups are structured appears to have implications for the nature of intra- and intersocietal relations.

Divale (1974, 1984), noting the correlation between matrilocality and external warfare, cast these variables into a causal relationship by linking the emergence of uterine kinship groups with territorial expansion and hostile campaigns against outsiders. He proposed that the so-called "matriliny problem" can be explained as an intermediate strategy that is employed by invading populations to break up their own fraternal interest groups, thereby reducing internal discord and mobilizing their male forces against a common enemy. This model relies on the premise that patrilineal descent, patrilocal residence, and internecine warfare constitute the *basic state* of preliterate societies.[3] Matrilocality and matrilineal descent groups, by this reckoning, are nothing more than a structural solution to migratory stress. This solution, Divale argues, is necessarily temporary since the organization it engenders is marred by "structural weaknesses." These deficits he describes as ". . . the inefficiency, inconvenience, and psychological stress that occurs when residence and authority are determined through two different lines. . ." (1984: 27). Ergo, as soon as an immigrant matrilineal society reaches equilibrium in the new territory and the selective pressure against fraternal interest groups subsides, the model predicts that males will again attempt to localize and internal hostilities will resume. A migrating society is thus predicted to pass through a succession of phases, beginning and ending with patrilocal and patrilineal organization, with intermediate matrilineal/bilateral and uxorilocal-avunculocal stages. Divale estimated that the entire cycle of residence and descent change takes about 1,800 years from the initial migration to complete.

To test his hypothesis, Divale (1984) designed a "cross-ethnohistorical" study of thirty-three societies, comparing their postmarital residential patterns and associated traits with data on the relative antiquity of their habitation at their present location (estimate of past migration date). Divale reported that the findings of this study statistically confirm his hypothesis that matrilineal-uxorilocal societies are more recent migrants to their areas (i.e., within the past 500 years) and that the remaining societies in his sample generally fall along the predicted gradient in terms of social organization *versus* estimated migration dates:

> . . . the principle of parsimony requires the reader who decides to reject this theory, that he or she provide a plausible rival *process*—not simply a variable—that can account for all the associations found in this study plus the very "curious" fact that societies who are very recent migrants will tend to have uxorilocal-patrilineal social structures, that slightly older migrants will tend to have uxorilocal-bilateral structures, that mi-

grants who are older than that will tend to have matrilineal-matrilocal structures, that migrants who are even older than that will tend to have avunculocal-matrilineal structures, and that the oldest migrants or indigenous societies will tend to have patrilocal-patrilineal social structures. (1984: 207)

Divale's challenge to produce a rival process that otherwise explains his statistical correlations is perhaps fair play, but goes beyond the scope of this book. Short of such an effort, however, it is legitimate to question a number of assumptions on which the migration theory relies and that may suggest alternative interpretations. First, a basic premise is that matrilocality, and by extension matriliny itself, is an interim form of social organization—a shapeshifter of sorts—on a continuum flanked on either side by asymmetrical patterns of residence and descent, with patriliny at both the beginning and end points. Divale admits, however, that there is no documentation to support the claim that patrilocality represents the basic social form of preliterate societies, i.e., the universal premigratory state from which matrilocality allegedly derives. Indeed, on the basis of extensive cross-cultural data, Murdock (1949: 190) argued categorically that not only were there no documented cases of direct transition of patrilineal to matrilineal descent, but that it could not occur. Divale's assumption of a priori patriliny for sample societies seems to be largely based on the theory of the male supremacy complex (Divale and Harris 1976), discussed in the previous chapter.

Also questionable are conclusions drawn about the patrilocal-virilocal nature of the oldest migrant societies in the study's sample. The Mbuti pygmies, for example, fall into the study's category of peoples that have allegedly returned to their original or indigenous state several millennia after their original migration. But is it realistic to assume that the social organization documented for this refugee population in the 1950s necessarily resembles that of their ancestors over four thousand years ago? Are there not other factors, in addition to past migration, that may account for their current sibless, bilateral, and virilocal social groups? Similarly, the utilization of averages for past migration dates masks the range of variation among Divale's sample societies both within and between the study's residence and descent categories. Thus, in the matrilineal-matrilocal category, the Veddas migration date is estimated at over 2,400 years ago, whereas other societies in this same group, such as the Timbira, migrated as recently as 146 years ago. Similarly, some sample patrilineal-patrilocal societies, such as the Gros Ventre and Kikuyu, who are supposedly representative of the oldest migrants, have occupied their current area for only 106 and 375 years, respectively. How, then, shall these outlier societies be interpreted? Are they coming or going?

Divale's migration model confirms some of the findings of earlier investigators that correlate residence and descent modes with variables such as the type and frequency of warfare, the presence of men's houses and large living floor areas, and the degree of local male exogamy. The model goes on, however, to cast these synchronic data into a diachronic continuum, pointing to the disruptive effects of migration as the primary catalyst for unidirectional changes over time. Thus, in theory, if society A migrates into the original territory of society B, each will progress through the predicted stages of social change commencing from the date of contact. But in the real world, things are not that simple. The nature of residence and descent systems cannot be predicted by a priori notions of primeval patriliny, nor by the length of time a population spends in the usurped territory of another. Arguably, the fundamental elements of social organization are not dictated by the length of time a society occupies foreign soil, but by the prevailing ecological circumstances at a given time in a given niche. What matters more than the migration itself, perhaps, is what happens after they get there—the types of resources that are or become available, the economic and social technologies that are developed and pursued for their acquisition and distribution, and the changing external forces that continue to impact and modify the local ecology. This goes well beyond a society's need to marshal or demobilize the troops for military campaigns.

Individual case studies suggest that additional factors are operative in the establishment and success of uterine kinship systems. For example, Snow (1995, 1996) utilized Divale's migration theory to explain matrilineal-matrilocal organization among the Iroquois. He hypothesized that the ancestors of the Iroquois migrated northward from the Pennsylvania Susquehanna River basin into already-occupied areas of New York State and southern Ontario around 500–900 AD. The prevailing residential pattern of these early migrants was assumed by Snow to be patrilocal, but to have transitioned to matrilocality shortly after their arrival in response to warfare with the indigenous population. This transition was proposed to have occurred rapidly after initial settlement and to have been marked by the appearance of multifamily residences (longhouses), horticulture, and compact villages.

Snow's hypothesis was later challenged by Hart (1999, 2000, 2001), who utilized archaeological evidence to support the alternative theory that Iroquois maize cultivation and matrilocal residence *co-evolved*, and slowly over time rather than suddenly as proposed by Snow. Large multifamily houses (one of Divale's markers of matrilocal residence), for example, do not appear until the twelfth and thirteenth centuries AD, or about 400 years after the initial Iroquois migration date. Moreover, rather than being a consequence of migration, Hart suggested that Iroquois matrilocality could have co-evolved in situ with maize cultivation as an adaptive subsistence strategy—one that optimized both the exchange of maize seed and the intra- and intergenerational transmission of horticultural technology. Utilizing a Darwinian model of maize genetic evolution, he concluded: "Matrilocal residence would have favored the perpetuation of agricultural management traditions, innovations, and favorable maize gene complexes under all conditions more strongly than would patrilocal or neolocal residence" (2000: 164).

Archaeological reconstructions typically assume that Iroquois social organization at the time of maize adoption consisted of dispersed patrilocal bands. This interpretation points to the absence of traditional Iroquois multifamily longhouses prior to about 1200 AD and to cross-cultural studies that have statistically linked matrilocality with large and patrilocality with comparatively smaller dwelling unit spaces.[4] Hart, however, leaves the door open to the possibility that ancestral premaize Iroquois populations were already matrilocal. As noted below, uterine kinship organization is compatible with subsistence economies based on fission-fusion cycles. Settlement patterns characterized by fluctuating periods of aggregation and dispersal may account for the smaller average dwelling unit sizes that predominate in the earlier Iroquois archaeological record.

In summary, the migration model argues that matriliny is *derived* from patriliny and that it emerges and is sustained only under the special circumstances of territorial expansion and external warfare. A similar theory has been advanced by Jones (2011), who argues that matrilineal organization serves as an adaptive mechanism for increasing internal solidarity in tribal societies that are engaged in territorial expansion and external hostilities along cultural frontiers. That is, in the absence of state-level bureaucracies, female philopatry creates cross-cutting ties that connect neighboring local communities within the tribe and facilitates their collective action against a common enemy. Jones joins Divale in portraying matriliny as a specialized organizational phase, one that provides certain political and military advantages to expansive tribal societies. The relationship of matriliny and external warfare is thus elevated from correlation to causation. Accordingly, such theories imply that there are no conditions in which matrilineal social organization may develop in situ and exist sui generis as a stable, stand-alone ecological adaptation.

Its appearance among both foraging and horticultural societies worldwide, however, and its particularly strong representation among Indians of the New World suggest that there are additional conditions that favor uterine organization and that define the matrilineal niche. As Divale and Jones suggest, a key to understanding societies organized on uterine kinship principles may well relate to the unique demographic and social consequences of dispersing related males and localizing female kin. The broader question, however, is this: If matrilocal societies are particularly adept at consolidating and mobilizing their male labor force for external military activity, what else are they good at? What additional political, productive, or other advantages accrue to this type of social organization?

Demographics, Resources, and Labor Supply

Consistent female philopatry results in the local outflux and influx of adult males of varying kin affiliations. This pattern generates the spatial segregation of males from their natal lineages or lineage segments in which they continue to exercise social responsibilities and leadership roles. Male dispersal, however, is counterbalanced in matrilocal societies by the maintenance of close geographical proximity among intermarrying matrilineages (Oberg 1955). Indeed, the dominant pattern among both matrilineal foragers and hoe farmers is for communities to be composed of several lineages, either on a seasonal or year-round basis.

Schneider (1961) noted the multilineage character of matrilineal societies and their resistance to internal segmentation. With uterine organization, kinship groups and their loci of male power and authority are segregated *by design*. Male authority heads are distributed among a number of intermarrying matrilineages, creating a greater likelihood of cooperative cross-kin relations and political alliance. As noted by Watson-Franke (1992: 478), what has been traditionally framed as a situation of male divided loyalties actually serves as a mechanism for fostering "male multi-connectedness."

This perspective is useful for understanding why matrilocal societies differ markedly from their patrilocal counterparts in the extent to which they actively compete with others in the same niche. Local groups that concentrate both resources and defensive-offensive power in the hands of related males are far more likely to engage in hostile confrontations along kinship lines than are those organized around female uterines. Matrilineal-matrilocal societies achieve internal political stability by dispersing rather than consolidating these potentially explosive clusters of male kinsmen, thereby creating cross-cutting alliances among component kin groups.

What, then, is the matrilineal niche? How and under what conditions does uterine organization optimize the acquisition and distribution of fitness-based resources to community members? Two conditions seem essential. First, the ratio of resources to population must be high, such that the productivity of a multilineage network exceeds that otherwise achievable by individual component lineages. Whether it be game, fisheries, dense food patches, or arable land available for swidden farming, resources within the homelands of the multilineage community must be communally accessible, be internally uncontested, and equal or exceed the population's collective needs. Second, the social organization must provide a stable and effective structure for labor mobilization, both for production and for protection of the collective resource base. In matrilineal societies, the dispersal of related males at marriage draws a number of kin groups into economic and political alliance. This makes available to the collective a labor supply significantly larger than that available to individual lineages. This strategy is especially adaptive for the exploitation of abundant resources on either a seasonal or permanent basis. Matriliny provides an integrative structure for multilineage production and distribution, thereby increasing efficiency, reducing risk, and enhancing the fitness of all community members.

Environmental settings containing abundant large game or aquatic resources appear to have favored matrilineal-matrilocal organization in the past. Exploitation of migratory herd animals was most successfully pursued with large communal drives rather than by individual or small-group stalking. This strategy required the assembly of a large labor force on a coordinated and cooperative basis. Similar manpower demands and multilineage cooperation were required for the efficient exploitation of annual fish runs or the hunting of large sea mammals. The cross-kin alliance networks characteristic of matrilineal-matrilocal foragers were especially effective in orchestrating labor for such communal production activities on either a temporary or year-round basis. Seasonal fission-fusion patterns of multilineage communities calibrated with resource availability were characteristic, although some settings, such as coastal areas, also supported permanent settlements in prehistoric times.

The Okanagan Indians of the Upper Columbia River provide a typical example of matrilineal-matrilocal foragers. According to observations made by Ross (1904) in 1810, their sociopolitical units consisted of several matrilocal lodges that alternatively nucleated and dispersed twice annually. During the mid-June to October salmon run on the Columbia River tributaries, the lodges came together to construct fish barriers, each of which was overseen by an elected superintendant who distributed fish each morning and settled disputes over allocations. Four specialized work groups that cross-cut matri-lodge memberships were formed, consisting of fishing and hunting parties for men and fish-curing and gathering parties for women. The intensive summer food collection efforts were directed at building winter food stockpiles for burial in caches, with each lodge receiving an allotment of fish, meat, vegetables, and fruit. At the end of the salmon run, the matri-lodges dispersed to hunt and gather for four to six weeks and then nucleated once again into a multilineage community at a prearranged location for the duration of the winter. Dispersal in the spring began the cycle anew.

Matrilineal-matrilocal organization also predominated among aboriginal South American Gran Chaco Indians, who aggregated their labor forces into sizeable composite bands for the communal hunting of abundant rhea and guanaco herds.[5] While this same mechanism was utilized in historic times for amalgamating multilineage communities into confederacies for external military campaigns, particularly after their acquisition of the horse, there is no evidence to support the hypothesis of prior patriliny among these peoples. The articulation of cross-kin multilineage communities for communal hunting was also common among Indians of the North American plains and likely represents a very ancient Paleolithic subsistence strategy (Driver 1961).

In each case, uterine kinship organization provided the structure for peaceful multilineage activity and the pooling of labor for collaborative subsistence endeavors. This feature may help to explain why male dominance in subsistence is commonplace among matrilineal-matrilocal hunter-gatherers. Just as uterine organization enlarges male labor supplies for the exploitation of resources, so it does for their protection. Competitive relations are reserved for external groups who threaten access of the collective to the abundant resources at hand.

Matrilineal-matrilocal hoe farmers, like foragers, are associated with habitats containing internally uncontested resources—notably, in this case, arable land and flora/fauna in which the multilineage community generally has an undivided interest. The frequent impermanence of gardening plots associated with swidden cultivation militates against the establishment of exclusive property rights and internal competition for land. Gardening activities are undertaken by localized groups of related women who, as noted by Hart for the Iroquois, may serve as the keepers and transmitters of embodied wealth in the form of crop management traditions and crop innovation. Males collaborate on a cross-kin basis in subsistence endeavors to supplement lineage food stores and for the protection of the collective resource base against external threat.

A major conclusion drawn by Gough (1961b) and Aberle (1961) is that matrilineal societies have a limited ecological range due to their concentration in extractive and hoe farming subsistence types and their association with the lower ranges of productivity. Perhaps more precisely stated, they are distributed over a broad latitudinal and geographical range, but primarily in niches that offer abundant resources that are most efficiently exploited with communal multilineage workforces over all or part of the year. As cultivators, they lack intensive farming techniques, large domesticated animals, and surplus economies. They are diverse in terms of the relative mobility, size, and complexity of their communities, but tend to be egalitarian in nature. Production is pursued primarily on a subsistence basis, and consequently the major forms of intergenerational resource transfer consist of embodied and relational wealth. In the more resource-rich habitats, such as the Pacific Northwest Coast or American Southwest, matrilineal societies developed larger seasonal or year-round communities, status inequalities, and some forms of material wealth. For example, rights in stationary or divisible resources, such as shellfish beds, salmon streams, or granaries, were in some cases held by specific groups or lineages, leading to status or wealth differences. Matrilineal societies with robust subsistence economies also sometimes developed higher levels of political complexity and centralized forms of governance, as exemplified by multilineage chiefdoms and confederacies.

In summary, matrilineal-matrilocal societies are associated with a signature complex of traits. They are particularly adept at creating and maintaining stable alliance networks between groups of kinsmen whose members achieve greater fitness through cross-kin cooperation and collaboration than through fission and competition. This unique balancing of lineal and affinal interests appears to be sustained by viable subsistence economies in which the benefits of such collaborative activity are equitably distributed and in which the primary forms of intergenerational resource transfer consist of embodied and relational wealth. As discussed below, the specificity of this niche type is further highlighted by the conditions under which uterine kinship groups become eroded and eventually disappear.

The Matrilineal Puzzle

Back in 1950, Richards coined the phrase "matrilineal puzzle" to refer to the asymmetrical allocation of authority over women and their children that matrilineal systems assign to brothers versus husbands. The perception has been that such systems of kinship reckoning and inheritance create conflicting loyalties for males, who must prioritize the interests of their sister's children (their matrilineal kin) over that of their own biological offspring. Richards went on to propose a number of "solutions" to the puzzle provided by the adoption of postmarital residence patterns that localize various combinations of cooperating lineal and affinal kin.

In his introductory essay to the classic volume *Matrilineal Kinship*, Schneider (1961) reinforced the matrilineal puzzle concept by maintaining that female-centered kinship systems create tensions and conflicts for in-marrying males because they have the following disruptive effects: downplaying or ignoring male procreative roles; militating against strong and stable husband-wife bonds; limiting male authority over their wives; preventing male access to jural rights in their children; and straining the relationship of fathers and their offspring. Implicit in this argument is the Western gender philosophy that matrilineal descent creates an "unnatural" arrangement between two sets of males who "naturally" seek to dominate the economic and reproductive lives of women—husbands in the domestic sphere and brothers-in-law in the sphere of the descent group.

Schlegel (1972) devoted an entire book, *Male Dominance and Female Autonomy*, to exploration of the matrilineal puzzle, a concept that she characterized at the beginning of her work as a product of ethnocentric befuddlement: "What may be a puzzle to the Western ethnographer is simply a fact of social organization for the native, no more and no less puzzling than any other fact of organization" (1972: 1). She conducted a comprehensive cross-cultural analysis of domestic authority and its correlates in matrilineal societies. The outcome was an authority gradient based on the degree to which a woman is subject to the dominance of her brother, her husband, or neither. Domestic group profiles were outlined for each of these types involving traits such as postmarital residence, marriage types, incest taboos, control of property, and female autonomy. But since this was a purely synchronic study, no attempt was made to explore historical or other factors that could shed light on the conditions underlying domestic authority patterns or how they resonate with authority in the public domain. We were thus left with static profiles of "solutions" to the matrilineal puzzle, in which women appear either as the passive spoils over which men compete for dominance and entitlements or, alternatively, as escapees from subordination through the forging of female alliances and the successful balancing of husband and brother roles.

Twenty years after Schlegel's classic study on matriliny and authority, some scholars questioned whether the alleged conflict between male lineal and affinal roles was overblown, or simply a product of ethnocentric perception. Watson-Franke (1992) revisited the question of the matrilineal puzzle by taking a fresh look at how masculinity is defined in matrilineal versus Western societies. The perception that matrilineal systems create inherent conflicts for males, she argued, is largely in the eyes of the beholder. The key difference between matrilineal systems and those steeped in patriarchal heritage lies in contrasting gender philosophies and the separation of male sexual and procreative rights.

Watson-Franke draws on Schlegel's earlier study to emphasize that matrilineal philosophies mitigate intersexual dominance by creating a cognitive orientation toward women rather than men. Motherhood and female generative powers are portrayed positively as sources of strength and influence. The autonomy of matrilineal women is also enhanced by their economic independence, which is vested in their descent group rather than their sexual partners. Instead of concentrating the socioeconomic authority of men in the roles of husband and father, matrilineal societies diffuse male rights and obligations across the multiple roles of father, husband, brother, and uncle:

> As sexual partners, on the one hand, men as husbands gain sexual access to women, though not juridicial rights to their own children; as brothers, they gain access to women's procreative but not their sexual potential; and as uncles they embrace their sisters's children as their own social and spiritual offspring and assume corresponding rights and obligations in the matrilineally defined descent group. (Watson-Franke 1992: 478)

Contrary to Schneider's (1961: 14) contention that the statuses of father and husband are not required in matrilineal societies, Watson-Franke maintains that matrilineal men play a multiplicity of roles within and outside the domestic unit, serving as nurturing mentors to their own sons and as enforcers of rules and obligations to their nephews. What are typically viewed by outsiders as "divided loyalties" among matrilineal men, she argued, actually serve as a prescribed mechanism for social cohesion between lineal and affinal kin. In other words, marriages forge cooperative relationships between kinship groups rather than individuals. A woman's brother and husband assume complementary rather than conflicting roles in this partnership. Watson-Franke notes:

> When compared to the monolithic nature of Western masculinity which is manifested in the exclusive male control of sexuality, matrilineal masculinity reflects in a kaleidoscopic view a diversity of male statuses which allow men to play influential but less controlling roles in various spheres

of their kingroups and communities by sharing, cooperating, and connecting. When we consider this gender philosophy the matrilineal puzzle dissolves. (1992: 486)

On this point, it is useful to return to Schlegel's (1972) original concept of a gradient of matrilineal domestic authority. Watson-Franke's model corresponds closely to Schlegel's Neither-Dominant type, in which matriliny is combined with matrilocal residence and female autonomy, including women's control over their own sexuality and property and their attainment of ascribed positions outside the domestic sphere. Historically, this is the aboriginal or symmetrical variety of matriliny, in which residence and descent, or natal and marital households, are aligned and in which the socioeconomic roles of brothers and husbands are well-defined and complementary. The pre-fur trade Innu and the seventeenth-century Iroquois are examples of this type. At the other ends of the authority scale are the Brother-Dominant and Husband-Dominant types, in which residence has shifted to patterns that favor the localization of related males, such as avunculocality and virilocality. This "masculinization" of matrilineal descent groups is typically accompanied by increasing male control over female sexuality, lineage property, and children, a greater potential for the ambiguous allocation of male authority, and a corresponding adjustment in the ideological portrayal and valuation of male and female roles.[6]

Recent investigations of the matrilineal puzzle have shifted focus from role conflicts to how males can pursue their inclusive fitness within the structure of a matrilineal system of inheritance.[7] Such studies generally accept the premise, either implicitly or explicitly, that matrilineal systems represent a deviation from a normative patrilineal state, in which male fitness is maximized by the linear transfer of wealth to their male offspring. The perception is that such investments are compromised in matrilineal societies by paternal uncertainty. This presents matrilineal males with the paradox of calculating, consciously or unconsciously, the odds of their mate's fidelity, and formulating strategies such as polygyny to, in essence, "beat the system." For example, a matrilineal male with multiple wives can profit by the transfer of inherited wealth to each of his offspring from their respective mother's brothers, rather than having to divide his own estate among some or all of his issue. Matriliny is thus proposed to evolve as an evolutionarily stable strategy under conditions where males can realize greatest reproductive returns from polygyny and diagonal rather than linear transfers of resources. This male inclusive fitness model thus proposes to solve the matrilineal puzzle and the origins of matriliny itself in a single stroke—as both an antidote to female promiscuity and a clever asymmetrical investment scheme by males in their own biological progeny. Two elements, however, have been traditionally neglected in such models: (1) female reproductive strategies and the way in which they intersect and interact with male strategies to affect the fitness of both sexes; and (2) the types of wealth available to males and females that contribute to reproductive success, the conditions affecting their transfer, and how the nature of transferrable wealth has changed over time.

As noted earlier, the introduction of heritable forms of material wealth and the participation of males in external market-based economies has served to undermine matrilineal systems of subsistence production and distribution and has been a key factor for social change in both the post-Neolithic and postcolonial periods. Rather than assuming, as did Schneider (1961), that matrilineal descent establishes structural relationships that are inherently conflicted, an alternative approach to understanding such systems is to identify the conditions under which the alignment of female natal and marital domestic units is disrupted or replaced by the localization of male kinsmen. When viewed in this context, the matrilineal puzzle is less an inevitable feature of matrilineal systems than it is a function of historic trends in the nature of political economies and the changing scope of intergenerational wealth transfer between parents and offspring.

Shennan (2011) utilizes the framework of niche construction to describe evolutionary trends in the nature of intergenerational wealth transfer and the way in which the increasing significance of material resources has impacted reproductive strategies and reproductive success. While hunter-gatherer societies and simple farmers generally have few material resources and are notoriously egalitarian, any adaptation that provides greater sedentism, population density, and the accumulation of surpluses has the potential to generate resources that are excludable and divisible, and subject to private ownership. When material property, such as land, livestock, currency, or embodied status, becomes a resource in intergenerational transfer, the potential for inequality emerges. This has important implications for inclusive fitness strategies. Inheritance of material wealth has a positive feedback effect on embodied and relational resources, enhancing not only the economic standing of beneficiaries, but their health, social status, and overall reproductive success. Notably, material resources and stationary property rights have intrinsic value, are often contested, and historically have been concentrated in the hands of males. Male role conflicts arise not from the inherent structure of matrilineal systems per se, but from changes in the nature of heritable wealth and their creative efforts to redirect these resources to their lineal rather than affinal kin.

The Vanishing Matrilineal Niche

Murdock (1949) was one of the first to propose types of residence, descent, and kinship terminology as dynamic systems. He recognized matrilocality to be particularly vulnerable to change in the face of new economic conditions that significantly influence the status or importance of males. A central factor for Murdock was the development of any form of moveable property or wealth, such as livestock, slaves, or money, that can be accumulated in quantity by men. Material wealth was seen as encouraging a transition from matrilineal to patrilineal inheritance and favoring the localization of male kinsmen. He observed that a matrilineal-matrilocal society under such stresses may transition to avunculocal or directly to patrilocal residence, and subsequently to patrilineal descent (i.e., through male purchase of jural rights in children).

Similar observations on factors affecting the integrity of matrilineal-matrilocal systems were made by Aberle (1961) and by Gough (1961b) in the classic book *Matrilineal Kinship* (Schneider and Gough 1961). In her analysis of matrilineal societies discussed in that volume, Gough drew attention to the effects of an external market system in which societies were often required to participate, but in which they had limited access or control:

> I suggest that certain specific characteristics of modern economic organization, found in varying degrees in all the societies we have discussed, bring about the disintegration of matrilineal descent groups. These are not, however, "trade" per se, or the use of money, or even the existence of markets... The root cause of modern kinship change in these societies appears rather to be the gradual incorporation of the society in a unitary market *system,* in which markets cease to be isolated and are linked in a common standard of value, and in which all produced goods, but more particularly land and other natural resources, and human labor itself, become privately owned and potentially marketable commodities. (1961b: 640)

Gough recognized that this external market system had disruptive effects on unilineal descent groups generally through the introduction of private ownership rights in land, wage labor, and a shift of primary production to external entities, such as factories or plantations. Such changes, however, had disproportionate impacts on matrilineal systems:

> Matrilineal groups seem to be badly hit as soon as their members enter the market system. Although they may not disintegrate altogether for many decades, they are likely to break down into their minimal segments. Further, as soon as individuals begin to acquire private earnings,

violent tensions occur between conjugal and paternal ties on the one hand and matrilineal ties on the other. Patrilineal groups seem better able to weather the early changes and may even give a temporary appearance of increased durability. (1961b: 649)

This greater resilience, Gough reasoned, stems from the fact that women in patrilineal systems are more fully incorporated into the descent groups of their spouses as mothers of heirs, thereby strengthening the elementary family. This contrasts with the situation of matriliny, in which the elementary family is torn between the descent groups of the husband and of the wife and children.

Although Gough's analysis was limited to farming societies, hunter-gatherers with uterine kinship groups were disrupted by external market systems in similar ways. As noted, matrilineal-matrilocal foraging societies in pre-Columbian and precolonial times were particularly well-represented in stable ecological settings with abundant resources, such as coastal areas, river systems, and woodland prairies that supported robust native flora and fauna. Because these were the same areas coveted by invading Europeans, matrilineal peoples were among the earliest and hardest hit by territorial displacement, depopulation, and acculturative influences. Historical records document specific attempts by missionaries, traders, and colonial bureaucrats to actively promote changes to matrilineal sexual practices, marital relations, postmarital residence patterns, and inheritance systems that clashed with their own cultural values and agendas. But, as Gough suggests, the primary threat to the survival of uterine kinship groups was posed by the disruption of aboriginal subsistence systems on which the integrity of their social networks was based.

The impact of foreign market systems on matrilineal foragers is illustrated by the postcontact experience of the Innu (Montagnais-Naskapi), referenced in the previous chapter (Leacock 1954, 1955). As recorded in Jesuit missionary accounts, original Innu socioeconomic units consisted of small matrilocal bands that dispersed during the winter and then nucleated during the summer months on the coast or at one of the larger lakes where food was plentiful. The Innu lived almost exclusively on the taking of abundant fish and game found within a loosely defined and commonly held territory. Following the introduction of the fur trade, however, Innu men were increasingly drawn into the foreign trapping economy, a shift that initiated fundamental changes in the structure of their society. Trapping gradually displaced communal large-game hunting as the dominant economic pursuit. The trading post replaced natural area subsistence hubs as the geographical focal point for summer band aggregation. There, furs were exchanged for major food supplies, with flour and lard overtaking meat as dietary staples. Material goods such as canvas and cloth replaced leather, formerly manufactured by women, for clothing, tents, and canoe covers. And importantly, the aboriginal pattern of band seasonal dispersal was exchanged for the setting of winter traplines, the maintenance of which fell primarily to males due to the burden of carrying supplies into the interior sufficient to feed larger (and now essentially nonproductive) family units. Several factors, then, converged to increase the individual economic importance of men in domestic units and to decrease the relevance of older cross-kin networks for communal subsistence activity. With the abandonment of winter band dispersal and increasing male competition for furs, winter traplines evolved into distinct territories that were inherited patrilineally. By the time Leacock conducted her fieldwork in the 1950s, band territories had developed where none had existed previously, and postmarital residence patterns were predominantly ambilocal, trending toward patrilocality.

Aberle's (1961) cross-cultural study of 565 societies presented in the *Matrilineal Kinship* volume reaches parallel conclusions regarding the conditions that typically undermine matrilineal-matrilocal organization. Not surprisingly, these conditions, such as the introduction of divisible and accumulative forms of wealth and male control of the primary forms of production, are the same as those discussed below that favor patrilineal organization. Aberle's famous line that "the cow is the enemy of matriliny, and the friend of patriliny" (1961: 680) has been supported by a number of studies on the spread of cattle herding to African horticultural societies. Murdock (1949: 373), for example, cited the case of the Herero, whose ancestors lived in sedentary matrilineal communities. Upon introduction of cattle to the region, they abandoned cultivation completely for an independent pastoral existence identical to that of neighboring Khoi-Khoi (Hottentot) herders. The pastoral conversion was accompanied by a shift to virilocal residence and the subsequent evolution of exogamous patrilineages alongside the older uterine descent groups.

The so-called "matrilineal belt" of eastern and southern Africa has attracted a number of ethnographic studies documenting the effects of the market economy on indigenous horticultural societies. In his studies of the Goba, Lancaster (1976, 1979) noted that agricultural land was traditionally regarded as a "free good." It is only when its value is upgraded from subsistence to more intensive production that greater male involvement, control, and a bias toward patriliny is favored. He observed that matrilineal descent groups have been weakened both by the loss of permanently cultivable lineage gardens and by growth of a local cash economy that has encouraged material wealth transfer from fathers to sons and patri-affiliation claims on children.

More recently, Holden et al. (2003), noting the positive correlation between the introduction of cattle and conversion to patriliny, linked changes in descent to the fitness benefits of inheritance rules. They argued that since the benefits of horticultural wealth are similar for sons and daughters, daughter-based inheritance and matriliny constitute an adaptive parental investment strategy where the risk of parental uncertainty in a son's children is high. In contrast, they suggest that any transferrable resources that benefit sons over daughters, such as cattle, increase male-biased parental investment. The accumulation of self-perpetuating cattle herds allows a son to enhance his reproductive success through the ability to support several wives, thereby promoting patrilineal inheritance and patrilineal descent.

Finally, there are rare examples where matrilineal institutions persist in the face of conditions, such as pastoralism, that would otherwise be expected to signal their demise. The Asben Tuareg of the African Sahara, for example, are reportedly matrilineal with avunculocal residence despite the fact that they have been camel herders for centuries. Some adaptive advantage, therefore, must accrue to uterine organization. As Murdock (1959: 408–9) explains, the Tuareg occupy the top of a complex interethnic caste hierarchy throughout the Sahara. This privileged position was apparently acquired by force and provides them with rights to tribute, serfs, and slaves from settled patrilineal peoples. In this case, matriliny serves to prevent the transmission of property to the offspring of intercaste marriages and hence maintains caste distinctions while broadening the range of alliances among the privileged minority.

Conditions Favoring Patrilineal Organization

Kinship groups formed on the principle of agnatic linkage clearly predominate among societies in the worldwide *Ethnographic Atlas* sample, with over 46 percent coded as patrilineal and nearly 70 percent as patrilocal/virilocal (Murdock 1967). Societies relying on male-centered kinship organization are not only more numerous than their matrilineal counterparts, but they have a much broader ecological range. They are represented among foragers and horticulturalists and prevail among pastoralists and intensive irrigation farmers as well. Moreover, they span a wide range of sociopolitical complexity, from simple bands and tribes to chiefdoms and highly centralized states. Indeed, it is the very pervasiveness of patrilineal organization in the ethnographic present that has led some theorists to propose genetic antecedents for human social forms. Historically, matriliny has been perceived as the exception that demands a special explanation, whereas patrilineal-patrilocal organization just *is*—the alleged manifestation of a universal human biogram.

The alternative position taken here is, once again, more in keeping with that of Hamilton, et al. (2007b). Namely, kinship systems are tools for niche construction that optimize the acquisition and distribution of fitness-related resources, such as energy, materials, genes, and information, for their members—male *and* female. The question then becomes: How do patrilineal-patrilocal systems differ from matrilineal-matrilocal systems in their handling and management of these resources? What are the distinctive features of the patrilineal niche?

Kinship and Subsistence Dominance

As with matrilineal origins, several theories have proposed that patriliny is the outcome of the relative contribution of the sexes to subsistence. A common theme in both evolutionary and sociobiological models is the association between agnatic kinship groups and the alleged food provider role of males. Thus, Steward (1955) and Service (1962) based their patrilineal-patrilocal band models on the assumed dietary importance and collaborative nature of male hunting. Cross-cultural studies of foraging societies, however, have failed to support this hypothesis (Martin 1974; Marlowe 2004). In fact, hallmark patrilocal bands, such as those of historic Australian aborigines, are often found in marginal or resource-poor habitats and tend to have the highest dietary dependence on female gathering.

Other investigators have proposed that male contributions to subsistence in horticultural societies will always assume primary importance unless male labor is otherwise diverted, such as for military activity. But attempts to statistically correlate female-farming systems with external warfare and male absenteeism have also been unsuccessful (Ember and Ember 1971; Sanday 1973, 1974). Women are frequently the primary cultivators among horticulturalists regardless of the prevailing nature of unilineal descent.

Where the productive importance of men and patrilineal kinship characteristically overlap, as among pastoralists and intensive farmers, male socioeconomic dominance is sometimes linked to the greater strength and mobility required for large animal husbandry and the attachment of livestock to the plow. Such explanations, however, ignore the parallel influence of portable wealth introduction and its intergenerational transfer on the nature of social groups. Defining the distinctive features of the patrilineal niche requires a closer look at the demographic and political consequences of male philopatry for resource acquisition and distribution.

Demographics, Resources, and Fitness

The most important contrast between uterine and agnatic kinship systems lies in their disparate rules for the spatial distribution of related males and females and the demographic and political consequences of this distribution for resource acquisition and management. Although males are recruited for authority roles regardless of the prevailing nature of descent or affiliation, virilocality brings the loci of power into alignment with residential or territorial groups. There is thus a greater likelihood of not only economic but political autonomy for component kinship units.

As discussed earlier, the creation of fraternal interest groups in patrilineal-patrilocal societies has been linked to a high incidence of feuding and internal warfare. Sahlins (1961) also noted the tendency of agnatic lineages to segment, thereby providing a mechanism for the spatial redistribution of local populations at each generation. Such atomization may be adaptive in marginal environments where lineage splintering and dispersal serves to lessen the effects of limited resource availability or depletion. Conversely, resource scarcity may also develop as a result of market forces. With market-induced scarcity, focus is on the competitive accumulation of divisible and heritable wealth that defines the boundaries not only of kin allegiance and conflict but of status and inequality. Where intergenerational material wealth transfer is limited to one or a small number of potential heirs, lineage segmentation provides an avenue for the establishment of new kinship units by excluded male offspring.

Due to their propensity for fission, agnatic multilineage alliance systems are more fragile than matrilineal-matrilocal ones. Female philopatry provides a built-in structure for cross-kin networks through the interlineage exchange of males and asymmetrical allocation of power and authority. Patrilineal-patrilocal systems, in contrast, must rely on the creation of other social or bureaucratic mechanisms for the cementing of broader cooperative networks. These may include the reciprocal interlineage exchange of women and associated forms of compensation (i.e., brideservice or brideprice), suprakin networks that ally autonomous patrilineages into clans, or pan-tribal sodalities such as age sets or other nonkinship groups whose memberships cross-cut the political allegiances of local fraternal interest groups.

The organization of kinship groups on patrilineal versus matrilineal principles reflects a fundamental difference in the way male and female fitness strategies are structured. With both types of organization, the respective reproductive goals of the sexes are the same. Male strategies are focused on access to fertile females as well as defense of the perimeter within which critical resources are found and on which the sustenance, security, and survival of their offspring rely. Female strategies focus on the successful nurturance of offspring through establishment of intra- and intersexual alliance networks that structure access to critical resources within this margin of safety. It has been argued here that matrilineal-matrilocal societies emerge primarily in noncompetitive situations where resources are plentiful and relatively uncontested. Under such environmental conditions, female philopatry enhances the fitness of both sexes by providing multiple mating opportunities, effective alloparenting networks (i.e., sororal interest groups) for the survival of resultant offspring, and a cooperative, articulating structure for the assembly of resource-efficient labor groups and defense of the collective resource base.

When critical resources are depleted or are more intensively exploited, however, large-scale cross-kin subsistence efforts are often less efficient or infeasible and the cooperative social networks on which they are based become dysfunctional. Under conditions of resource scarcity, the inventory of resources available to kinsmen is more finite, demarcated, and contested. The perimeter of male resource protection shrinks accordingly, from a multilineage to a lineage or extended family scale. In such cases, the kin group architecture becomes masculinized, favoring localization of male uterines and eventually male agnates. Male productive and defensive efforts are refocused from affinal to lineal kin networks, and subsistence and military objectives become closely aligned.

For females, sororal interest groups for both food getting and alloparenting are disrupted by asymmetrical residence patterns that separate them from their kinswomen. Females must therefore pursue fitness strategies for themselves and their offspring more independently, with greater reliance on the protection and goodwill of consorts and affines. Accordingly, in the absence of uterine subsistence and alloparenting networks that ensure the nurturance of young, male fitness relies on greater oversight and control of in-marrying females and their offspring. As producers of both food and potential heirs, women in patrilineal societies may themselves be regarded as a resource or form of wealth over which men compete. In more complex societies, females often leave direct production entirely and become publicly cloistered by males who provide for their offspring. In patrilineal societies with market-induced forms of resource scarcity, the fitness of both sexes becomes importantly related to the ability to accumulate and transfer material wealth to their male offspring.

Although male and female fitness strategies are structured differently in matrilineal and patrilineal systems, they represent flexible and complementary responses to environmental conditions rather than biological imprinting. As observed by Poewe (1979), the principles by which uterine and agnatic kin groups are organized also have consequences for male-female relations and for differential access of the sexes to power and status:

> Matrilineal descent, combined with the cultural idealization of emergence from a womb, results in the general recognition that everyone derives ultimately from a common womb. It gives rise to an egalitarian ideology—an equality, no less, between all men and women, between all *wombmates*. Different fathers do not matter. . . . Patrilineal descent, combined with the evidence of deriving from a specifiable womb, implies a differentiation between individuals. Excepting siblings of the same mother, everyone derives from a *different* and *discrete* womb since descent ties through women are ignored. From the perspective of women, alliance takes ideological precedence over descent. Relationships are nonegalitarian, and the nature of one's status depends on the position of the father and on the relationship of one's mother to one's father. While different fathers are incidental in matriliny, in patriliny mothers always matter, even if only as indices of discreteness. In matriliny, mothers have the effect of uniting and equating; in patriliny they have the effect of separating and distinguishing. (1979: 116)

Scarcity, Wealth, and Productivity

It is proposed here that patriliny and the localization of male agnates confer an adaptive advantage in situations where fitness-related resources are contested, whether these resources are in short supply due to natural conditions or whether, in more favorable environmental settings, their control is seen as critical to the accumulation and transmission of material wealth.

In subsistence-based societies such as hunter-gatherers, male-centered kinship groups are expected where access to native flora and fauna is limited or where the ratio of resources to population is low during the major portion of the annual cycle. Under these conditions, survival may be enhanced by the proliferation of small, autonomous patrilocal bands that maintain an ecological balance through limited-scale extractive endeavors. Collaborative subsistence activity at the supraband level is less productive in these societies due to both resource limitations and the challenges of sustaining alliances among multiple fraternal interest groups for mutual benefit purposes.

Unlike their matrilineal counterparts, however, patrilocal bands have the option to alternatively initiate and terminate alliance networks in response to ecological conditions without altering their core memberships. For example, clans evolved among related patrilocal hunting bands of interior Canada as an articulating mechanism for brief summer nucleation in the Upper Great Lakes and among the equestrian Puelche and Tehuelche Indians of South America for warfare and large-scale hunting endeavors in postcontact times. In each case, band amalgamation made larger labor supplies available for the exploitation of seasonally available resources. Because patrilocal bands are economically and politically self-contained, however, cross-kin alliances were tenuous and brittle. Such bands tended to ally with one another only if their respective interests were promoted and to terminate such unions if these interests were in any way compromised. Puelche bands, for instance, united only for warfare against a common enemy. Similarly, Tehuelche bands, although joined by their recognition of a common leader for external warfare, did not hesitate to launch attacks against one another for territorial (i.e., resource control) violations (Steward 1946: 150–52, 164).

The more typical portrait of the patrilineal-patrilocal band drawn by Steward and Service was one of small, nomadic, and dispersed extended families—autonomous agnatic groups that occupied discrete territories containing sparse resources. While both authors proposed patrilocal bands as models of Paleolithic life, ethnohistorical data suggest that such societies in modern times represent the remnants of refugee populations that suffered habitat loss through natural events or that were displaced and driven into marginal areas by more complex populations. The ancestors of aboriginal Australians, for instance, once occupied the shores of ancient Pleistocene interior lakes that fell victim to climate change. Centuries later during the historic period aboriginal peoples were slaughtered and driven into the arid interior from dense settlements on resource-rich coastal rivers (Martin 1970). Given the mix of matrilineal and patrilineal elements surviving in the complex Australian marriage class systems, it is likely that their precontact organization deviated sharply from that of modern aborigines.

The social organization of horticultural societies is of particular interest, since they exhibit the greatest diversity of male-female involvement in subsistence activities and the strongest correlations with specific unilineal descent and residence patterns. Previous cross-cultural studies have found that while female farming prevails among hoe farmers generally, productivity levels vary with the nature of social organization (Martin and Voorhies 1975: 213–41). Where local groups are organized around a core of female uterines (matrilineal-uxorilocal), the relative dietary importance of cultigens is low. In contrast, where local groups are formed around male uterines (matrilineal-avunculocal), relative productivity rises sharply, and again when local groups consist of male agnates (patrilineal-virilocal).

Among hoe farmers with male localization, females no longer cultivate communally with their own kinswomen, but individually for their respective spouses and affinal kinsmen. The concomitant increase in relative productivity accompanying this change in labor groups was attributed by Boserup (1970) to the development of polygyny, a central feature of avunculocal and virilocal farmers regardless of the prevailing nature of descent. Boserup viewed polygyny as a mechanism for the economic development of simple farming communities in the absence of intensive techniques. The combined efforts of several females on behalf of an individual male household head greatly increases the yield of cultigens available for consumption and surplus exchange, along with male control over their labor and productive output. Women in polygynous households assume a dual role as producers of cultigens and producers of children and thus themselves become particularly valuable as units of exchange among groups of male kinsmen.

In habitats with large tracts of arable land, there has been an evolutionary trend toward population expansion, increased productivity in cultivation, and heightened competition for resources and surplus wealth. These conditions parallel those of poor environments in that the increasingly negative ratio of resources to population stimulates both competition and an orientation toward maximal production. Both of these factors select against stable matrilineal-matrilocal forms and initiate a trend toward male localization and patrilineal descent. Surplus hoe farmers utilize the greater military potential of male localization for resource competition and balance the divisive tendencies of agnatic groups with more complex forms of political integration. Productive and distributive activities become greatly elaborated and not uncommonly dichotomized on the basis of sex—females being largely responsible for subsistence and surplus production and males for the manipulation of exchange networks. The more closely crop production becomes linked to economic competition over surplus goods themselves, the more likely that male labor will become directly involved.

Trends among hoe farmers toward the masculinization of primary subsistence activities, of kinship groups, and of distribution and exchange with increased productivity are accelerated with the development of plow and irrigation agriculture (Aberle 1961; Gough 1961b; Martin and Voorhies 1975: 276–300). The invention and gradual spread of intensive farming techniques is likely attributable to hoe cultivators geared toward maximal production and the exchange of surpluses. By far the majority pattern among kin-based intensive cultivators is patriliny and virilocality, along with a fundamental shift in the sexual division of labor to male farming systems. Patrilineal intensive cultivators base not only their redistribution and exchange groups around related males, but their labor groups as well. This systematic exclusion of women from food production effects basic changes in the nature of domestic groups. Polygyny, for example, becomes obsolete. With males as the primary cultivators, the accumulation of multiple wives is an economic liability rather than an asset. While optional or limited polygyny may be preserved as a legitimate indicator of prestige and economic success, monogamy generally emerges as the dominant pattern, along with a general decrease in the size and complexity of domestic units. Large patrilineally extended families give way to minimally extended, stem, or nuclear family units whose male heads share land rights in common.

In accounting for the overwhelming dominance of male farming among intensive cultivators, historical factors must be taken into account. That is, although the elimination of females from food production arose initially as an adaptive response to new modes of production, notions on the propriety of this labor division (and the domestic isolation of women) accompanied the diffusion of intensive techniques themselves. Much of the consistency in economic, social, and political role playing among the sexes observed in contemporary agricultural societies lies in the fact that intensive cultivation evolved independently in very limited areas of the Old and New Worlds and that these innovations were spread in both ancient and modern times by cultures with expansive economies. Nowhere, for instance, do we find the introduction of the plow without accompanying biases on how and by whom it shall be used. Unlike simple hoe cultivation, which may have enjoyed many independent innovations over a broad ecological range, intensive farming is associated with a specific complex of culture traits that developed in a single kind of niche and diffused as a package, often by force, to new areas.

Among intensive farmers, men almost universally assume primary responsibility for the production of cultigens, whereas women orient themselves around the complementary activities of processing raw produce in the domestic context. Surpluses are increased by more efficient technologies and are amassed in considerable quantity on a seasonal basis, as opposed to the more piecemeal returns of year-round hoe farming production. This not only changes the distribution patterns of domestic groups, but also makes available more concentrated revenues for pan-kinship (i.e., political or bureaucratic) distributive organs. With intensive cultivation, an increasing dichotomy emerges between rural *subsistence* economies based on kinship and urban *exchange* economies based on the articulation of individuals, goods, and services in nonkinship institutions according to the market principle. Whereas centralized political economies based on hoe cultivation rely primarily on the redistribution of internal surpluses, those based on intensive farming place principal reliance on effective participation in external markets and the replacement of large corporate kinship groups with more efficient units of production.

Finally, societies that derive their primary subsistence from animal herding are almost universally associated with patrilineal-virilocal kinship groups. Segmentary lineage organization and a high degree of militarism are also commonplace. As suggested by Murdock (1949), male economic dominance is likely to prevail in societies relying on animal husbandry, where herds constitute a form of self-generating and accumulative wealth, and where the pasturage or other animal welfare requirements dictate the structure and movements of kin communities. Pastoralism is generally presumed to derive from mixed farming populations. As a secondary adaptation, it has served as a safety valve for the expansion of overburdened or highly competitive sedentary communities. This heritage is illustrated by the preference among former hoe and intensive farmers for polygyny and monogamy, respectively. Similarly, where pastoralists engage in supplementary cultivation, they tend to mimic the sexual labor division patterns of the horticultural or intensive agricultural communities from which they originated (Martin and Voorhies 1975: 333–52).

Conditions Favoring Bilateral Organization

According to an early cross-cultural study of 250 societies conducted by Murdock (1949), approximately 70 percent were found to recognize some form of unilineal descent (62 percent) or double descent (7 percent), while the remaining 30 percent reckoned maternal and paternal kinship affiliation bilaterally. Notably, although they lacked distinct unilineal descent groups, 70 percent of the bilateral societies had preferred unilocal postmarital residence patterns, suggesting that such practices may represent vestigial remnants of former matrilineal or patrilineal kinship organization. Similarly, the fact that the remaining 30 percent of these bilateral societies favored ambilocal or neolocal residence may be indicative of societal changes associated with urbanism and the emergence of nuclear families.

Subsequent worldwide samples have indicated that the relative frequency of bilateral kinship systems is over 35 percent (Murdock 1967). If, as has been argued here, kinship systems are social technologies that serve to optimize the acquisition and distribution of fitness-related resources, how is the representation of bilateral kinship systems across a broad spectrum of societies, from very simple to very complex, to be explained? And where does bilaterality fit in terms of human evolutionary development?

Bilaterality as Primeval

As discussed throughout this book, considerable academic energy has been devoted to musings on human social origins. American and British schools of thought became convinced early on that Paleolithic social life was built around clusters of related males by virtue of their alleged economic and political dominance. Patrilineal and patrilocal band models prevailed as prototypes of Stone Age society in twentieth-century theory and were supported by initial ethnographic descriptions of preagricultural peoples. Subsequent field studies, however, often bolstered by ethnohistorical reconstructions, revealed that the social organization of hunter-gatherers was not as uniform and male-centered as previously assumed. On closer examination, it was found that about 75 percent of contemporary foraging societies were not only bilateral in descent, but exercised multiple options for residence after marriage (Martin 1974; Marlowe 2004).

Sociobiological theory has recently embraced the notion of bilateral descent as a prototype for primeval human society, albeit as an intermediate or backdoor route to male philopatry and agnation. That is, while the biological kinship link between mothers and offspring is treated as obvious and undeniable, such theories require a vehicle by which paternal kinship could be recognized and ultimately prevail in ancient society. Bilaterality provides such a bridge, one allegedly constructed by males through enforced female dispersal, polygynous or monogamous pair bonding, and paternal investment in offspring. Contemporary bilateral foragers, however, exemplify neither the male socioeconomic dominance nor the parasitic intersexual relationships envisioned by some sociobiological models of primeval society.

The notion that bilaterality was the universal platform from which unilineal descent groups evolved was also summarily rejected by Murdock (1949: 219). Ecological explanations of why bilaterality predominates among foragers point instead to the limited resource base available in more marginal habitats, the sizeable geographic range over which resources are distributed, and the extremely low human population densities. As Murdock noted, such conditions favor the formation of small, highly mobile social groups or kindreds whose members maintain ties with a wide variety of maternal and paternal relatives. Since kindreds are formed on the basis of contemporary egos, their memberships and boundaries are highly flexible. Ambilocal residence patterns represent an opportunistic strategy in that individuals may choose to reside with a broad spectrum of relatives or other partners at various points in their lifetimes based on the benefits that may be locally available.

The proposal that such bilateral kindreds formed the basis of primeval social organization begs the question of what Pleistocene environmental conditions were like. Given the choice, would our ancestors have preferred to live in habitats such as the Kalahari Desert or the Mediterranean Coast? The American Great Basin or the Columbia River Valley? Most would agree that well-watered, resource-rich environments would have been favored over arid, desolate ones. The ethnohistorical record has provided glimpses of Paleolithic adaptations in such prime habitats—ones that offered abundant food and the potential for robust unilineal descent groups and regional alliance systems. There were undoubtedly times in the Pleistocene when the effects of climate change in certain areas of the world challenged our ancestors to live in less favorable habitats—ones whose ratio of resources to population was low. Under such conditions, ancient peoples, like modern hunter-gatherers, would have been forced to splinter into smaller kinship or family groups to sustain the basic needs of life. It is erroneous, however, to assume that such conditions of scarcity predominated in the past. A closer examination of foraging societies in recent years has brought into question the traditional patrilineal-patrilocal band model and other unitary prototypes of primeval social life. Instead, it has highlighted the sheer variety of potential adaptations that likely evolved in diverse and dynamic Pleistocene landscapes.

If contemporary hunter-gatherers commonly lack unilineal descent and localized, philopatric kin communities, then explanations for their absence must be sought in the conditions that both support and undermine the viability of corporate consanguineal kinship groups themselves.

Bilaterality as Devolved Uniliny

As noted, the majority of contemporary hunter-gatherers, though bilateral, continue to recognize preferred unilocal residence rules, suggesting that such practices may have been more consistently observed in the past. Indeed, there is increasing ethnohistorical evidence to indicate that these remnant foraging populations were formerly organized into discrete unilineal descent groups—social forms that were systematically unraveled by a combination of severe depopulation, territorial displacement, and the diminution of native flora and fauna.

Service (1962) was one of the first to propose that foragers organized into bilateral composite bands or small family groups represented degenerative types of social organization resulting from the disruptive effects of European contact. The decimation of South American hunter-gatherer societies through epidemic diseases is a case in point (Gusinde 1931, 1961; Steward 1946: 56–57, 83; Martin 1969). For example, the Ona of Tierra del Fuego were originally organized into thirty-nine patrilineages that were grouped into two distinct and mutually antagonistic regions of the island. Their population, estimated as at least 2,000 in 1875, was reduced to 300 in 1910, 100 in 1935, and 50 in 1944. The earliest ethnographic observations of the Ona were made in 1918–1925 by Gusinde, when they were already decimated. The neighboring Alacaluf and Yahgan suffered a similar fate. At first contact the Alacaluf were numbered at "a few thousand," but had been diminished to 160–200 individuals by 1945. Similarly, population figures for the Yahgan were estimated at 2,500–3,000 around 1875, but their numbers fell to 1,000 by 1884, 400 by 1886, 200 by 1889, 130 by 1902, under 100 by 1913, and to only 40 by 1933. Under such circumstances, unilineal kinship groups become dysfunctional as resource management units due to the enormous and rapid loss of personnel, forcing survivors to fission into small bilateral family groups.

In other parts of the world, such as Australia and Africa, hunter-gatherers were pushed into increasingly marginal areas by more complex populations that usurped the most favorable habitats for themselves. Aboriginal Australians living in resource-rich areas were not only deliberately exterminated en masse by incoming European colonists, but the native flora and fauna on which surviving groups relied was severely disrupted by large-scale sheep herding. Similarly, African foragers such as the desert Bushmen and forest pygmies were territorially displaced by a succession of farming and herding populations in recent centuries, forcing them into habitats with limited resources. In each case, localized unilineal descent groups, with their multigenerational memberships and collective production and distribution systems, could no longer be supported and atrophied into smaller, more efficient atomistic units.

In defining the conditions that favor bilateral descent, Western scholars need look no further than their own culture history in the post-Neolithic period. As tribal economies were drawn into and eventually subsumed by state political economies, the basic units of production and distribution were fundamentally changed. Although matrilineal and patrilineal systems survived the plow and irrigation revolutions to varying degrees, they became increasingly antithetical to the economic requirements of organized states, whether developed in situ or imposed from abroad. Since domestic production provided the lifelines of the urban market economy, rural surpluses and eventually real property itself came under the more direct control of bureaucratic agents. The stripping of redistributive functions from kin leaders destroyed the material basis of larger communal groups, forcing their members into cooperative ventures on a more limited scale, and individual ventures into nonkinship production groups. Participation in production groups external to the kin collective often favored the displacement of old unilocal residence patterns with neolocality, a factor that Murdock (1949: 208–9) notes as contributing to the disruption of existing unilineal groups and emergence of bilateral descent. Over time, unilineal descent groups atrophy into bilateral kindreds and ultimately, in the more urbanized centers, into autonomous nuclear families. As discussed earlier, a parallel process affecting the nature and persistence of unilineal kinship groups in non-Western societies has taken place in the European colonial period where external market systems compete with and eventually displace kin-based subsistence economies.

In summary, whether in very simple or very complex societies, unilineal kinship systems disappear because the productive and distributive systems on which they are based become dysfunctional or obsolete.

Kinship in Evolutionary Perspective

This chapter has approached kinship systems as tools for optimizing the procurement and management of fitness-related resources by and for members of a breeding population. This interpretation rests on three basic premises. First, individual fitness both relies on and is enhanced by group membership—by the reproductive opportunities and beneficial resource allocations that participation in group life provides. Second, the formation of human groups from ancient to historic times has rested on the basic primate pattern of sex-biased philopatry. The phenotypic expression of philopatric rules is not fixed, but varies with environmental conditions and the selective advantages it provides. And third, the way in which philopatric groups are structured in a given niche is a reflection of the ratio of critical resources to population, the degree to which these resources are accessible or contested, and the strategies by which they are most efficiently exploited and secured.

This rather neutral platform for understanding kinship systems proposes neither monotypic human biograms nor universal evolutionary stages. It acknowledges the significance of sex differences and synergies for the achievement of both reproductive and productive success and the ability of human groups to manipulate and structure these traits and propensities for mutual benefit. It is this very aptitude for behavioral plasticity that set genus *Homo* on a separate evolutionary trajectory. Neurobiological evidence suggests that sex-biased philopatry has played a central role in gene-culture co-evolution from the dawn of humanity. This same idiom of exclusive kinship has persisted over the millennia as a primary adaptive technology, its various forms shaped by the natural selective forces of diverse environments.

Theoretically, utilization of uterine linkage and utilization of agnatic linkage have been equal options as building blocks for food getting, food sharing, and mating groups since the origin of society. But, as we have seen, unilineal-unilocal variants have different demographic and structural consequences and hence potentially different adaptive advantages. Matrilineal-matrilocal kinship groups, for example, do not appear as independent isolates, but rather are combined into clusters of such units that share the resources of a given geographical area in common. The regulatory mechanisms of uterine systems are unique in that the special interests of component lineages or corporate groups are subordinated to the interests of the multilineage collective. Internal competition is effectively checked through the exchange of males who are then jointly responsible for protection of the collective from external threat. This organization allows for the efficient and harmonious exploitation of a given habitat by a sizeable population, the component units of which benefit from uncontested resources, joint subsistence labor, and joint protection. Such a system is adaptive only in those situations where resources meet or exceed the productive needs of the population in question.

The regulatory mechanisms of patrilineal-patrilocal kinship groups are contrastive. Agnatic lineages have the full complement of productive groups, distributive groups, and clusters of male kinsmen to pursue individual lineage interests on a competitive basis. It is this potential for independent economic and political action that gives patrilineal organization its greater ecological range. Economic groups based on agnatic affiliation may exist as local isolates in habitats with scarce resources or join into larger collectives for the more intensive exploitation of favorable environments. In the first case, volatility promotes survival. In the latter, the ubiquitous presence of multiple centers of power and economic interest must be met with new regulatory mechanisms that bind members of the collective into some real or fictive (i.e., clan or state) relationship. Patrilineal organization is thus adaptive where competition over either scarce or abundant resources is intense—the very conditions that undermine the basis of uterine kin group alliance.

It has been argued here that symmetrical matrilineal and patrilineal systems not only yield substantially different economic and political groups, but that sex-biased philopatry also provides a cultural means to maximize material success in a given niche in the absence of technological advance. In evolutionary terms, some general trends are observed in the way in which kinship systems articulate with resource acquisition, labor division, and productivity.

In societies geared toward subsistence production only, both matrilineal and patrilineal options are found depending on the nature of ecological adaptations. Labor division by sex shows great consistency. With the exception of foragers at extreme latitudes, women and men share responsibilities for the provision of dietary staples. However, since male activity in all societies follows the locus of economic competition, men also assume a major role in the manipulation of regulatory mechanisms. In societies with subsistence production only, the status quo to be maintained is the claim of a population to the resources of a given habitat from which dietary staples are derived. The ratio of population to resources will determine the degree of competition necessary for survival and hence the most adaptive distribution of males on the basis of kin criteria.

The movement of a society from a purely subsistence to a surplus orientation initiates a process of increased male involvement, not only in regulatory activity, but in the control of production itself. Some of the best examples we have for this kind of economic development are documented cases of descent change from subsistence-oriented matrilineal-matrilocal societies to surplus-oriented patrilineal-patrilocal ones. As matrilineal hoe farmers begin to exploit their environment on a more intensive basis, perhaps at the impetus of population pressure or external markets, the locus of competitive relations shifts dramatically. Whereas before, the collective shared resources in the face of external threat, now the economic interests of individual kin groups begin to consolidate in opposition to one another. Old regulatory mechanisms for control of internal competition, namely, the nonlocal distribution of related males, become increasingly antithetical to new productive goals. Matrilocal residence, therefore, is gradually supplanted by the spatial aggregation of male uterines. Matrilineal-avunculocal societies thus take on many of the same characteristics as patrilineal ones; namely, male localization, the use of polygyny for increased production, and a heightened frequency of competition and hostile confrontation along kinship lines. In the final phase of this transition, patrilocality arises as an option, often via the mechanism of material compensation to uterine heirs in lieu of inheritance rights. Patrilocal descent groups rise alongside uterine ones, their eventual dominance being dependent on the transfer of property rights to the consolidating core of male agnates.

The entire process of descent change in kin-based communities is thus much more than a social phenomenon. As the locus of competition shifts from the equitable distribution and protection of resources to the accumulation and exchange of produce itself, male regulatory activity becomes particularized along kinship lines and expanded to the control of production and distribution. The extreme of this developmental trend, namely, the entry of males into actual food production, is characteristic of hoe farmers with the highest cultigen yields.

Surplus production, while favoring the masculinization of economic institutions, also requires the development of new regulatory mechanisms for the maintenance of internal harmony among competing and politically volatile kin groups. Surplus wealth makes possible the transfer of regulatory functions to special-purpose bodies which, by privilege or force, coordinate productive and distributive relations within the boundaries of the sociopolitical unit, as well as cooperative and competitive relations abroad. This transfer of both responsibility and power is particularly essential to the development of the state and to the attrition of unilineal descent groups. What begins as an intensification of productivity in the traditional economy eventually generates a dual economy—one system based on the production, distribution, and surplus accumulation of goods within landed kinship groups, and the other based on the exchange of all factors of production (raw materials, and eventually labor, land, and other forms of capital) according to the principles of supply and demand. The productive dimension of the new economic order encompasses a percentage of grassroots labor (taxation, military, and corvée obligations), involuntary labor (slavery), gainful manipulation of capital on the open market, and technological innovations (i.e., metallurgy for tools and weaponry, irrigation systems, etc.) designed to increase existing revenues. With economic development, dimensions of the new order proliferate into a number of specialized institutions for, among other things, the organization and control of internal power, free market labor, capital flow, raw material acquisition, and external trade networks.

Throughout the entire process of growth, the new economic order gains at the expense of the old, utilizing technological innovation and capital gain to liberate an expanding segment of the population from direct food production and to redirect their labor into the manufacture of market goods. The grassroots economy, along with the labor divisions and groups formed on the basis of unilineal kinship, are thus doomed to envelopment by a system that draws resources, labor, rewards, and allegiances into the legal fiction of states and anonymous corporations.

When viewing the long pathway of human evolution from Plio-Pleistocene to historic times, trends in technological development, productivity, population density, and sociopolitical complexity have tracked with the rise and demise of unilineal descent systems. Past and present theorists have linked specific kinship prototypes to points on this evolutionary continuum based on gene-drive or sexual dominance. Others have entangled the agnostic process of natural selection with notions of inevitable progress or political morality. Such theories err by disregarding not only the diversity, but the *niche-based* distribution of kinship types through time and space.

It is expected that matrilineal, patrilineal, and bilateral organization were all represented among Stone Age foragers, depending on the selective advantages they provided in specific environmental settings. It is also expected that changes in the frequencies of kinship types over time is related to trends in the nature and availability of compatible ecological niches. Thus, for example, niches favoring matrilineal organization, namely, those with abundant and uncontested resources, are assumed to have been commonplace in the past, but to have declined precipitously in the face of competition and changes in the nature and availability of critical resources. The expansion of patrilineal organization over time is attributable to its compatibility with the growing number of niches in which resource scarcity or competitive surplus production predominates. Similarly, the death knell of unilineal descent groups themselves was sounded by a niche change where older tribal systems of production and distribution were supplanted by urban market economies.

The transfer of regulatory and eventually productive and distributive functions to nonkinship groups in surplus societies has made philopatric alliance and the cultural embellishment of sex differences less and less relevant to survival. Consumption units have gradually eroded from kin collectives to the basic reproductive dyad, and corporate production groups have moved, however slowly, toward recruitment of labor on the basis of assiduity, skill, and creative potential rather than on the basis of sex. Observed evolutionary trends are the story of how humans, armed with certain material and social technologies, have moved from niche to niche over the millennia. New material and social technologies are now global in scope and will take the species to the unprecedented ground of pursuing productive and reproductive success in an increasingly nonkinship world.

ns="http://www.w3.org/1999/xhtml">
#  Epilogue

My original intent in writing this book was to explore the question of why human societies, for the bulk of our evolutionary history, have organized themselves into social groups based on uterine or agnatic principles. Previous research had convinced me that, as among nonhuman primates, male and female philopatry have distinct demographic, economic, and political consequences that correlate with ecological variables. Anthropological interest in kinship theory has waned in recent years, however, and the question of matrilineal and patrilineal antecedents has itself been rendered moot by monotypic evolutionary models. A popular if not prevailing theory is that agnatic linkage not only is a universal organizing principle, but has biological or genetic roots in our ancient primate past. In pursuing my original mission, therefore, it was necessary to step back and examine the evidence on which this conclusion is based, a task that resulted in a fundamental refocusing of the present work. What began as a cross-cultural examination of unilineal kinship systems morphed into to a cross-disciplinary journey on human social origins.

As it turned out, peeling back the layers of nested hypotheses that support the theory of ancestral male kinship was somewhat reminiscent of deconstructing a matryoshka doll. Removing the outermost nested layer revealed the core hypothesis that anisogamous reproduction and innate dominance dictate the primacy of male reproductive strategies for shaping primeval hominin social life. Beneath the second layer was the corollary hypothesis that female reproductive success relied on the establishment of pair-bonded relationships or sexual contracts with males for family provisioning and protection. Under the third nested layer was the premise that such provisioning emerged when males coalesced into bands of brothers and engaged in cooperative subsistence endeavors, invented tools and fire, and supplied sources of protein critical to both group survival and cortical expansion. And venturing down to the final layer of the matryoshka, we find the hypothesis that climate-based deforestation and the expansion of grasslands during the Late Pliocene epoch served as the primary impetus for arboreal abandonment by early hominins and their newfound reliance on the hunting of mammalian herbivores. Primeval habitats are painted with unpredictable or widely scattered resources that limit the size and densities of ancestral human groups. Here, aridity, scarcity, resource competition, and intergroup aggression are prevailing conditions of Paleolithic life, driven by extreme climate volatility. Taken together, this set of nested hypotheses is sometimes referred to as the "standard narrative" of human evolution, the layered components of which are presented as so pivotal to the survival and reproductive success of ancient humans that they were genetically imprinted as the indelible biogram of our genus.

One of the fun things about matryoshkas is the careful way in which they are crafted, such that all the layers fit nicely together. This is certainly the case for the standard evolutionary narrative, which presents a parsimonious explanation of human social origins based on a set of interlocking hypotheses. But, in the end, the narrative is still a story and parsimony does not make it true. This book has endeavored to disassemble these layers, examine the assumptions on which each is based, consider more recent evidence, and develop alternative hypotheses that may lead to novel reconstructions. In short, there's more than one way to craft a matryoshka.

This book has invited the reader to essentially start the layering process over, this time from the bottom up. It has been argued that the traditional savanna hypothesis concept of open, semi-arid grasslands is being challenged by evidence for mosaic Plio-Pleistocene biomes. Detailed paleoecological studies at early hominin fossil sites are painting a much different picture that includes well-watered and resource-rich habitats, some of which may have served as perennial refugia during climate extremes. These new findings suggest that the importance of lakes, rivers, deltas, estuaries, and coastlines as ancient occupation areas during the gradual deforestation phase of post-Messinian-Event Africa has been greatly underestimated. While the focus of human evolution has been on the interlacustrine region of the East African Rift Plateau, areas that now comprise the Saharan and Kalahari deserts were also once covered with lakes and waterways that were home to early bipedal apes. The secrets buried under these now vast barren expanses may one day yield new fossil evidence that further revises our portrait of early hominin life. There is a general reluctance, it seems, to move away from the scorched-earth portrait of Plio-Pleistocene landscapes and to acknowledge the ability of our ancestors to successfully pursue and occupy waterside habitats at or near the forest fringe.

The nature of Plio-Pleistocene ecology has profound significance for the next nested layer, namely, ancient hominin subsistence patterns. It is now generally acknowledged that the earliest adaptations to terrestrial life were not based on the cooperative hunting of mammalian herbivores on the open plain. While there is evidence for the scavenging of carnivore kills, game meat now appears to have constituted an unreliable and relatively minor dietary component throughout the Early Pleistocene. In contrast, occupation of biodiverse mosaic habitats suggests the potential for intensive gathering of a wide range of flora and fauna, including aquatic species. The presence of dense food patches could potentially support sizeable groups on a seasonal or year-round basis, significantly altering the stereotypic portrait of early hominins as small, wide-ranging nomadic bands. Direct evidence has also been noted for the ancient procurement of fish, a critical source of protein and dietary lipids. This book has maintained that *co-reliance* on aquatic and terrestrial fauna throughout the Pleistocene has been generally overlooked. Gathering additional evidence on dietary breadth is critical for reconstructing the social demography, density, and migration patterns of ancient hominin groups, as well as the nutritional basis for encephalization. The hunting hypothesis, although recently discredited, remains a positive cultural image and a persistent Paleolithic legend.

The domino effect of hypothesis nesting is particularly evident as one moves up to succeeding layers of this alternative matryoshka. With the removal of collaborative hunting as the mainstay of early hominin subsistence, the pivotal role played by male provisioning to pair-bonded units is effectively begged. It has been suggested instead that the mammalian primate pattern of female gathering and provisioning of young is likely to have extended to proto- and early human terrestrial life as well. This book has supported the hypothesis that the clustering of matricentric family units into bonded alloparenting groups provided the foundation for intergenerational care and food sharing and for the essential group energetics and life history changes that shaped the evolution of the genus. Ancient female hominins within such groups forged mutually beneficial alliances with males as sexual partners, helpmates, and defenders of group interests. But there is nothing to suggest that such intersexual relationships were founded on parasitism, economic dependency, or the nuclear family dyad.

Moving up to the next nested layer of the matryoshka, one confronts the cornerstones of male and female reproductive success. In lieu of coercion, male provisioning, and pair bonding, how was early hominin fitness achieved? Neurobiological research findings have been presented that suggest that the emancipation of reproductive behaviors from limbic system control and the capacity for behavioral plasticity were critical to the emergence of convergent fitness strategies among the sexes. The process responsible for reproductive emancipation was mosaic brain evolution, in which the various systems of the brain were reorganized in a manner that provided greater connectivity, cognitive function, and cortical control, along with the capacity for flexible and nuanced behaviors. As comparative neurological studies of bonobos and chimpanzees have demonstrated, this capacity is marked by actual brain differences in neuronal types and densities and in neural pathways that dramatically impact sociality and reproductive behaviors. These findings serve as a further caution against viewing contemporary apes such as chimpanzees as avatars of our last common ancestor. The bias maintained throughout this book is that human males and females did not evolve as predators of one another's interests—hominins began from a different place. What distinguished hominin lineages was the ability to realize enhanced fitness through the forging of cooperative intra- and intersexual alliances. The exercise of these mutually beneficial strategies most likely evolved within the intensely social setting of stable multimale-multifemale groups rather than through infrequent encounters between members of scattered nomadic bands.

Multimale-multifemale groups are commonplace among contemporary primates occupying niches with discrete patches of high-quality foods, a resource distribution pattern that may have also typified the more favorable mosaic Plio-Pliocene landscapes in which hominins now appear to have evolved. Such groups are also correlated with female philopatry and with permanent female-bonded kinship groups, a type of social organization that is compatible with the emergence of allocare networks. Recent studies have additionally suggested a genetic link between this type of social organization and the process of mosaic brain evolution. Namely, female imprinted genes appear to have played a critical role in the expansion of cortical function, the modulation of limbic system responses, and the evolution of behavioral plasticity. Critical changes in hominin energetics may have also been transmitted through the vehicle of mitochondrial DNA mutations in the female genome. As noted by researchers, the frequency and transmission of such changes at the molecular level would have been accelerated in breeding populations and gene pools in which related females remained together in their natal group.

Building on the foundation of these alternative nested hypotheses, what does the ancestral hominin social group look like at the final or uppermost layer of our reconstructed matryoshka? It is a cohesive community made up of both sexes, the core of which consists of a number of permanent and probably ranked female-bonded kinship groups, the female members of which remain with their natal group during their lifetimes and the adult male members of which disperse at adolescence. Females collaborate with their kinswomen in food getting, food sharing, and childcare endeavors. Adult males gain membership and attach themselves to such non-natal communities by forging cooperative intersexual alliances for polygynandrous mating, defensive or protective activities, and occasional food sharing. Subsistence is derived from the intensive gathering of a variety of terrestrial and aquatic flora and fauna. These resources are at least seasonally abundant and distributed in dense, defensible patches in proximity to water sources. Fission-fusion patterns are commonplace, with communities segmenting along kinship lines during the leanest portions of the year and nucleating into sizeable, even multicommunity gatherings at times of unusual faunal abundance (i.e., cyclical fish runs or migrations of terrestrial mammals). Ancestral hominin groups occupying such prime habitats are prosperous, well-nourished, and fecund, and serve as source populations for geographical expansion when resources decline or when density otherwise exceeds local carrying capacities.

In short, by peeling back, re-evaluating, and reassembling the set of nested hypotheses on which the standard evolutionary narrative is based, it is possible to arrive at an entirely different conclusion. Although arguably no more speculative than its alternative, a new Plio-Pleistocene model that breaks rank with innate dominance, male philopatry, and wandering hunting bands may be viewed as inconceivable by some, heresy by others. Theories of ancient human social life centered around female kinship and egalitarian sexual relationships may be welcomed in other quarters as a boon to feminist doctrine, a reaffirmation of the goddess, or the mirror of precapitalist society. In fact, they are none of these.

The reason that an ancestral female kin group hypothesis deserves serious consideration is that it fits the biological, genetic, and ecological facts very well. As mammals, female hominins were the natural providers and nurturers, and bearers of the heaviest burdens of evolutionary development, including the increased energy demands of pregnancy, birthing, lactation, and extended care that accompanied encephalization. The genus could not have broken through the gray ceiling without the emergence of alloparenting, to which female kin were especially attuned. Similarly, the life history changes that marked our evolutionary pathway relied, in large part, on mosaic brain evolution, in which the female genome played a major role. A reasonable argument can be made that the convergence of fitness strategies enabled by reproductive emancipation and the capacity for behavioral plasticity was necessarily negotiated in the socially challenging context of multimale-multifemale groups, a type of primate social organization linked to female-bonded kin groups. The size and permanency of such groups could only have been maintained in premium habitats with abundant resources. Such environmental settings, although not spatially or temporally dominant, *did* exist in several regions of the African continent during critical periods of the Plio-Pleistocene. It was perhaps the serendipitous and unprecedented convergence of all of these neurological, genetic, and ecological elements that allowed for the emergence of genus *Homo*.

If humans initially evolved in social groups where female kinship units were central, does this support the old argument for a matrilineal stage in human evolution? Not really. What it *does* support is the notion that the sexes had different roles to play in the sequence of genetic and epigenetic changes that separated us from other ancient apes. As mammals, females were the sex in the bull's-eye of critical energetic and life history changes. Female fitness strategies and the female genome may therefore have had an oversized influence on the structure of early social groups. This is not to say that such changes could not have occurred, or did not occur, with other patterns of philopatry, only that female philopatry may have had the effect of accelerating these changes in the emergent hominin gene pool. Equally important, it seems, to this sequence of evolutionary events is a set of compatible ecological circumstances that allowed them to unfold. Resource-rich habitats that permitted males and females to aggregate into stable communities and to interact socially on an intensive basis were perhaps critical to the progress of mosaic brain evolution. Notably, these are the same types of ecological circumstances, namely, habitats with dense food patches or relatively uncontested resources, that favor female-bonded kinship among contemporary nonhuman primates as well as matrilineal systems among humans. As ancient hominins expanded their ecological range, they also expanded their repertoire of social forms, alternating their fitness strategies and philopatric rules in characteristic ways to optimize energy capture in relation to available resources.

A central theme of this book is that behavioral plasticity allowed hominins to become the architects of their own reproductive success—by modulating their mating behaviors for convergent benefit and by structuring their social groups in a manner that maximized energy recovery in a given niche. The epigenetic rules or social DNA for managing the two most basic elements of fitness—sex and food—have been limited in number and responsive to ecological conditions in consistent ways. The basic philopatric and mating rules that structure the distribution of females and males in relation to resources are shared with other primates and are therefore part of our ancient heritage. What set hominins apart some 5 million years ago was their increasing ability to manipulate and tailor these rules in response to stochastic conditions, an ability that became indelibly imprinted in the structure of our brains.

This is the legacy of modern humans, one that slowly evolved in the Pleistocene over millions of years. The post-Neolithic age, however, dramatically altered the ecological circumstances of human existence over what, in geologic time, was the mere blink of an eye. The social DNA that served us so well in the past was rapidly outpaced by the impacts of advances in material technology. Intensive food production, increased population densities, and the rise of centralized states escalated regional and global resource competition and inequities. The growing predominance of male philopatry among human populations in the past ten thousand years, often masqueraded as a genetic mandate, is instead a reflection of unidirectional changes in human ecology. Ancient philopatric rules for the regulation of mating and the formation of productive and distributive groups have become skewed or rendered obsolete, only to be replaced by anonymous institutions and virtual realities.

Homo sapiens sapiens, like other contemporary apes, has arguably become a species with its back to the wall, frequently reverting to a show of canines. It is reasonable to assume that natural selective processes impacting hominin brain reorganization continue to the present day, shaping new refinements that enable our species to further modulate primitive limbic behaviors and respond flexibly and creatively to conditions critical to survival. But natural selection keeps its own clock and plays no favorites. Only time will tell if our social intelligence is able to keep pace with our cognitive genius.

 Endnotes

Introduction. Some Givens

1. See Latour and Strum (1986) for a review and analysis of the shortcomings of human social origins theories.
2. See also James (2017). In recent years, British scholars have issued an appeal to social anthropologists to join the conversation and lend their expertise to the questions of human kinship and language origins. Such was the focus of a 2005 Early Human Kinship workshop sponsored by Royal Anthropological Institute (London), in concert with the seven-year British Academy Centenary Project (2003–2010) entitled "From Lucy to Language: The Archaeology of the Social Brain." For contributions emanating from these meetings, see Allen et al. (2008) and Power, Finnegan, and Callan (2017).
3. For detailed critiques of kin selection theory, see Nowak, Tarriata, and Wilson (2010), Nowak and Highfield (2011), and Allen, Nowak, and Wilson (2013), also republished as an Appendix in Wilson (2014). See also de Vladar and Szathmary (2017).
4. In the final chapter of *The Extended Phenotype*, Dawkins acknowledges that most of the book has been devoted to "playing down the importance of the individual organism, and to building up an alternative image of a turmoil of selfish replicators, battling for their own survival at the expense of their alleles, reaching unimpeded through individual body walls and with each other without regard to organismal boundaries" (1982: 250). Importantly, however, his bird's eye view of the natural selection process also recognizes the potential ways in which phenotypic behaviors such as cooperation and niche construction may influence the direction and success of genetic replication. Perhaps what Dawkins's critics find culturally objectionable is the prospect that such behaviors are guided by an invisible genomic hand, apart from individual or group volition.
5. The term "hominin" has recently replaced "hominid" in new classifications. Hominins include modern humans as well as the australopithecines and fossil members of the genus *Homo*. Hominids (members of the more inclusive family Hominidae) include hominins along with contemporary apes such as the gorilla, chimpanzee, bonobo, and orangutan.
6. Recently uncovered paleoarchaeological remains from the Jebel Irhoud site in Morocco, North Africa, dated as 300 ka, have been identified as belonging to *Homo sapiens sapiens*. These finds suggest a much earlier horizon for the evolution of our species than previously assumed, along with the possibility that modern humans evolved in multiple areas of the African continent. See Stringer and Galway-Witham (2017) and Gibbons (2017).
7. See especially the "Preface" discussion in Finlayson (2014: vii–xvii).

8. Monogamy refers to a single male-female mated pair. Polygyny refers to one male with multiple female mates, polyandry to one female with multiple male mates, and polygynandry to mating systems in which both males and females have multiple mates.
9. For more detail on the hunting hypothesis, see Washburn (1961) and Lee and Devore (1968). The patrilocal band concept was popularized by Service (1962). The hunting hypothesis model carried the corollary assumption that the camaraderie and spoils of the hunt earmarked males as creators of and providers to the domestic unit. The relative contribution of hunting by males to the diet of hunter-gatherers has long been questioned by Martin and Voorhies (1975: 178–83) and others. By the end of the century the notion of hunter-gatherer primary dependence on meat had been discredited by O'Connell et al. (2002). But this hypothesis lives on for followers of Tiger's (1969) male-bonding theories and in Wrangham's (2009) linkage of domestic unit formation to meat and cooking, and is arguably implicit in Wilson's (1975, 2012, 2014) concept of providers to the "nest."
10. For a comprehensive sociobiological perspective on the elements and evolution of human nature, see Wilson (1978).
11. See, for example, Dart (1953), Ardrey (1966), Tiger and Fox (1971), and Wrangham and Peterson (1996).
12. See Berger et al. (2015) and Berger and Hawkes (2017). The paradox presented by this ancient hominin, *Homo naledi*, is twofold: (1) the combination of a chimpanzee-sized brain with potential evidence for deliberate burial practices, and (2) the surprising results of recent dating for these remains, which places them as contemporaries with much bigger-brained and apparently more culturally advanced hominins.
13. "Anthropodenial" is a term coined by de Waal (2016: 22), which refers to the tendency to reserve certain traits, such as advanced cognitive abilities, exclusively for *Homo sapiens sapiens*.

Chapter 1. Perspectives on Anisogamy

1. New research suggests that eggs may play an even greater role in conception than previously assumed. Recent experiments have indicated that the combining of gametes at fertilization is not always random, leading to the hypothesis that sperm cells with unsuitable genes may be screened out by an as yet undetermined process (Nadeau 2017). See also Arnold (2017).
2. For a more thorough discussion of nineteenth-century sexual stereotypes and the persistence of sexual bias in twentieth-century science, see Martin and Voorhies (1975: 146–55), Hrdy and Williams (1983), and Landau (1984).
3. See, for example, Debetz (1961: 145–46).
4. A paper emanating from the 1966 international Man the Hunter symposium at the University of Chicago even suggested that the economic burden of allegedly nonproductive "excess females" in early human society was a reasonable explanation for the practice of female infanticide (Washburn and Lancaster 1968: 302).

5. The male-bonding hypothesis is most closely associated with Chance (1961), Fox (1967, 1972), Tiger (1969, 1970), and Tiger and Fox (1971).
6. The debate stirred by publication of Wilson's *Sociobiology* led to the formation by opponents of the Sociobiology Study Group and the convening of a 1977 symposium by the American Association for the Advancement of Science, the results of which were published in 1980 (see Barlow and Silverberg 1980). More recently the debate has focused on the validity of kin selection versus multilevel selection theories (see Allen, Nowak, and Wilson 2013).
7. For a detailed discussion and critique of these theories, see Small (1993: 191–202).
8. See Martin and Voorhies (1975), Slocum (1975), and Lancaster (1976).

Chapter 2. First Families

1. The Family Hominindae (and the term "hominid") includes modern humans, ancient bipedal apes, as well as contemporary apes such as the chimpanzee and gorilla. The term "hominin" is now reserved for a more exclusive subfamily group that includes humans, ancient members of the genus *Homo* (i.e., *habilis, erectus, heidelbergensis*), as well as the australopithecines.
2. See especially Lovejoy (2009).
3. For a critique of these and other evolutionary theories on female sexuality and mating behavior, see Small (1993: 191–202) and Hrdy (1984, 2000).
4. "K-Selection" refers to environmental conditions that favor breeding strategies that produce small numbers of slowly developing offspring, resulting in stable populations of long-lived individuals.
5. For a listing of monogamous primate species, along with a discussion of conditions favoring monogamy, see Hrdy (1981: 34–58).
6. DNA analyses of fossil materials have also been recently offered by other researchers in support of the theory of ancient male philopatry. Lalueza-Fox et al. (2011) conducted genetic analyses of Neanderthal fossil remains found at a cave site in Spain, dated 49 ka. The materials in question are teeth and bone fragments thought to represent the remains of twelve individuals (six adults, three adolescents, two juveniles, and one infant) that fell victim to a cave-in. Morphological observations and DNA testing were conducted in an attempt to determine their sexes and genetic relationships. This effort, however, was hampered by the paucity of *in situ* bone material and was only partially successful. Of the twelve individuals, sex determination was made on only nine (six male and three female) on the basis of tooth morphology, and only four (all male) on the presence of a Y chromosome. Testing for mtDNA lineages revealed that, with the exception of one individual, all belonged to one of two lineages (seven in one, four in the other). The authors concluded: ". . . it seems reasonable to assume further that these two lineages represent two groups of very close relatives in the female line" (2011: 251). However, based on the fact that the three

individuals thought to be female (based on morphology) carried different haplotypes, and that three of the four genetically typed males were of the same mtDNA lineage, the researchers concluded that "patrilocal mating behavior" was indicated. The authors attempted to bolster this interpretation by their observation that patrilocality dominates in modern societies and is associated with greater local diversity in female than male mtDNA lineages. However, no Y-chromosome information is provided that would genetically link the identified males patrilineally. Further, the localization of males closely related in the maternal line would also occur with the pattern of avunculocal residence. This is not to say that patrilocal residence was not practiced by some groups of Neanderthals, but that given the admitted uncertainty of sex-typing by dental morphology and the lack of genetic sex-typing for two-thirds of this small sample, such conclusions appear premature.

7. See Wilson (1975, 2012, 2014), Sherman et al. (1995), and Foster and Ratnieks (2005).
8. Jones (2000) proposes a parallel concept, namely, "group nepotism," which emphasizes the importance of "group-beneficial values" in intergroup selection.
9. As noted by Hrdy (1981), female primates actively seek multiple sexual partners as a means of establishing social networks that secure their group status and the safety of their offspring. Polyamorous liaisons are the norm irrespective of the prevailing formal mating structure. In a study of Japanese macaques, for example, nearly 30 percent of infants were found to be sired by males outside the polygynous group.
10. This notion contrasts sharply with prevailing assumptions of Service (1962), Wilson (2012), Boaz (1997), and Finlayson (2014), along with advocates of the male ancestral kin hypotheses, all of which characterize early hominin social groups as small nomadic bands pursuing a marginal existence in increasingly arid savanna habitats.
11. See, for example, Narroll (1965) and Mace and Pagel (1994).

Chapter 3. Paleoecology and Emergence of Genus *Homo*

1. Proto-human apes are thought to have radiated widely from the region of modern Ethiopia to southern and western Africa and to adjacent areas of Eurasia sometime after 3.5 ma. Eastern Africa is also most commonly pointed to as the geographical origin of the larger, bigger-brained *H. erectus* around 2 ma. Fossils of similar antiquity appear in Southeast Asia, but no proto-human forms thought to be ancestral to *H. erectus* have been found in this area. For a summary of major theories on early hominin evolution, see Maslin, Schultz, and Trauth (2015).
2. For papers arising out of these conferences, see Washburn (1961) and Washburn and Lancaster (1968).
3. For a comprehensive evaluation of the hunting hypothesis, see O'Connell et al. (1999).

4. Arguments for the potential use of Acheulean handaxes as tools for hunting or predation have been reviewed and largely debunked in a paper by Whittaker and McCall (2001). Recently, investigations by Wilkins et al. (2012) of stone tools recovered at the 500 ka archaeological site of Kathu Pan 1 in South Africa have reported smaller retouched points that could have been hafted and utilized as spear tips. But, to date, there is no evidence linking earlier *H. erectus* populations to stone weaponry or big-game hunting.
5. See also Binford (1981) and Kline (1987).
6. See, for example, Langdon (1997) and Lowenstein and Zihlman (1980).
7. See Vaneechouette, Kuliukas, and Verhaegen (2011).
8. The importance of the aquatic food chain as an essential source of DHA has been challenged by Carlson and Kingston (2007), who argue that sufficient DHA can be produced by humans through the consumption of wild terrestrial foods containing Omega-3 fatty acids. They also claim that fishing occurred too late in human evolutionary history to be a factor in brain size increases. Cunnane et al. (2007) rebutted these arguments on both grounds, citing data indicating that human DHA production is insufficient to maintain normal infant brain development, and citing evidence from the fossil record linking fish consumption to early hominin sites.
9. For evidence of *Homo erectus* fire use, see Alperzon-Afil (2008), Beaumont (2011), and Gibbons (2007).
10. The early fossil record seemed to indicate that increases in *H. erectus* body size were markedly greater for females than males. See, for example, McHenry (1994) and O'Connell et al. (1999). This asymmetry has been linked to factors such as nutritional sufficiency, broader subsistence ranges, and the reduction of female energy loads attributable to alloparenting and infant altriciality (Anton, Potts, and Aiello 2014). Other explanations suggest the possibility that ecological changes may have a greater effect on female than male biomass (Gordon, Johnson, and Louis 2013), or that pregnant and lactating females may have had differential or priority access over males to the highest-quality food patches (Hrdy 1981). Still other investigators have proposed that increases in female body size may have been selected for simply because larger females produce larger babies, produce a greater quantity and quality of milk, and are better equipped physically to compete with others for access to preferred foods (Ralls 1976). However, because of the polymorphic nature of pre- and early *Homo* forms, greater clarity on sex differences in a variety of morphological traits must await a more complete inventory of fossil specimens.
11. For an expanded discussion of the "grandmother hypothesis," see Hawkes, O'Connell, and Blurton Jones (1997), Hawkes et al. (1998), Aiello and Key (2002), Opie and Power (2008), and Hrdy (2009).

Chapter 4. Paleolithic Dinner Pairings: Red or White?

1. While this study found bone marrow to have significantly lower bacterial counts than carrion meat, making it safer to consume, the investigators

note that under natural conditions cross-contamination would likely result due to contact of the marrow with nonsterile tools and with exterior surfaces of the bone containing blood, meat, and other tissue residues.
2. For theories on *H. erectus* migration routes, see Gabunia et al. (2000), Vekua et al. (2002), Ferring et al. (2011), and Gibert et al. (2016).

Chapter 5. Signature Hominin Traits

1. For a useful summary of the evolutionary impact of cooked food, see also Gibbons (2007).
2. In their assessment of the caloric intake and nutritional content of ancient foraging diets, Ryan and Jetha (2010: 175) reference the conclusions of Harris (1989) and Diamond (1997) that the emergence of agriculture resulted in a marked decline in human dietary sufficiency and overall quality of life. See also Harari (2011).
3. See, for example, Malinowski (1956, orig. 1931).
4. See Ryan and Jetha (2010: 213–62). Lovejoy (2009), for one, argues that early hominins were not polygynandrous because human sperm, unlike that of primates with a high degree of male sexual competition, lacks a coagulum. This substance forms a vaginal plug after ejaculation that serves to block the penetration of sperm from subsequent conspecifics. He suggests, on this basis, that ancestral monogamy among hominins was more likely. Ryan and Jetha (2010: 228–35) argue instead that sperm competition does indeed exist among human males, and that it involves two mechanisms that take place inside the woman's body both before and after ejaculation. First, the pre-ejaculatory thrusting of the uniquely configured human penis creates a vacuum that serves to draw out competing sperm in proximity to the ovum. Just prior to ejaculation, the head of the penis shrinks, breaking the vacuum and allowing the new spermatozoa to penetrate. And second, the initial and final spurts of ejaculate contain different chemicals—in the initial, chemicals that protect the spermatozoa from attack (from the female host or from the ejaculate of prior competing males), and in the final, chemicals that act as a spermicidal agent to slow the progress of both latecomers and newcomers. This suggests to the authors that hominin males can still play the competitive genomic game without developing (or retaining) the morphologically dimorphic armor required by physical confrontations in the external world.
5. Notably, while the incidence of infanticide by males is often linked to their desire to eliminate the genes of competing conspecifics from the gene pool, it is equally plausible that such violent acts are simply linked to the desire to accelerate sexual access to obviously lactating (and nonestrus, nonreceptive) females. Similar motives for expediting reproductive opportunities are noted by Kostjens (2008), who sees infanticide as a way of mitigating the limited breeding tenure of dominant males. Anyone who has housed a pair of breeding cats takes caution to isolate the tom from his newborn offspring, lest they become lunch. A parsimonious interpretation of such acts

is that the tom's efforts are programmed to speed along his next mating opportunity. Similarly, while infanticide is not uncommon among chimpanzee males, the fact that they are unable to recognize their own offspring complicates, it would seem, the argument for genetic cleansing. Perhaps the more critical factor in the frequency of infanticide is the extent to which the mothers of infants are sexually available or have been so in the past. In other words, a female's familiarity and establishment of friendly, polygynandrous relationships with proximate males over time may have more weight in ensuring the survival of her infant than its actual biological parentage. Among bonobos, where sexual opportunities abound for all, infanticide is apparently unknown.

6. The opposite of plasticity is "canalization." Canalization refers to the presence of robust phenotypes that are relatively impervious to environmental change and therefore may inhibit evolution unless "shocked" by novel conditions. Canalization can mask "cryptic" (unexpressed) genetic variation that may be released and lead to phenotypic variation if environmental stasis is disrupted. See, for example, Flatt (2005), Buskirk and Steiner (2009), Wright et al. (2010), and Moczek et al. (2011).

7. See Richards, Bossdorf, and Pigliucci (2010). New behavioral genomics research by Robinson and Barron (2017) suggests that instincts can evolve epigenetically. That is, where plasticity conveys an advantage to individuals that exhibit the trait early in their development or with less practice, instincts may be produced by natural selective forces that act to adjust the timing and intensity of epiallelic expression. The underlying theory is that learned behavior and instincts are regulated by the same neural mechanisms.

8. Notably, Harari (2011: 4) refers to presapient hominins as animals of no significance: "The most important thing to know about prehistoric humans is that they were insignificant animals with no more impact on their environment than gorillas, fireflies, or jellyfish."

9. Hybridization among Neanderthals and AMHs is now indisputable. Their ability to interbreed and produce fertile offspring also negates the theory that they belonged to distinct species. Interestingly, in this regard, my own DNA contains over three hundred Neanderthal genetic markers.

10. de Waal (2016: 175–76) notes in retrospect that while the "Machiavellian" label has been useful in describing the more competitive or tit-for-tat aspects of chimpanzee behavior, it fails to accommodate or advance an appreciation and understanding of the full range of their social skills, including cooperation and empathy. See also Byrne and Whitten (1988).

11. Still other studies have linked neocortex size to variables such as dietary and mating patterns. See, for example, Sawaguchi (1992), Pawlowskil et al. (1998), and Schillaci (2008).

12. The role played by von Economo neurons in human self-awareness, empathy, and social functioning is supported by ongoing medical research connecting various neuropsychiatric disorders, such as autism and specific forms of dementia, to damage to or destruction of these brain cells in the ACC and insular cortex (Chen 2009).

13. The finding that behavioral differences between chimpanzees and bonobos are correlated with actual brain differences suggests that their neurobehavioral specializations may reflect distinct evolutionary adaptations over the past 2 million years. Boaz (1997: 75–77, 80–81), for example, notes that bonobos, along with several other species currently living in the nonrefuge forested region south of the Zaire River, are not endemic species, but appear to have migrated in from other adjacent regions in response to past climate change events. Proposed candidate areas for their original homelands include East African savanna woodlands and forested patches along the Atlantic Coast. It is therefore possible that the bonobo distinctive behaviors may represent vestigial features of a social organization that evolved in more mosaic habitats.
14. As noted by Wengrow and Graeber (2015), the movement of human hunter-gatherers between groups of varying sizes and densities in response to seasonality or resource availability is an ancient Paleolithic pattern. While their emphasis is on AMHs, similar fission-fusion strategies may have also been characteristic of more ancient hominins, whose group sizes and associated social alliance networks extended beyond immediate kin to include membership in intentional multilineage communities.
15. Dunbar (2009) later adjusted his estimates for optimum group size among hominins and, in a subsequent joint paper (Gowlett, Gamble, and Dunbar 2012: 695–96) noted that the social brain hypothesis "on a broader and deeper level" addresses the relationship between neocortex size and the relative complexity of social cognition, rather than group size per se. While the coauthors recognize the potential for human communities to achieve significant size through the development of scalar networks, they still support Dunbar's Number as the optimal size beyond which social groups become stressed and inherently unstable due to neocortical limits and constraints.
16. Fission-fusion social organization has also been observed among elephants and several species of carnivores, ungulates, and even fish. See Lehman, Korstjens, and Dunbar (2007); Amici, Aureli, and Call (2008).
17. Chen and Li (2001) studied genomic divergences among hominoids as a means of estimating effective population size. They concluded that the effective population size of our last common ancestor was five to nine times greater than that of the subsequent human lineage. Such higher densities at their point of origin would be consistent with the geographic expansion of these populations throughout the Old World in the Early Pleistocene. The authors hypothesize that the reduction in population size may reflect subsequent local extinctions and recolonizations.
18. Interestingly, general systems theory could be applied in a parallel manner to mosaic brain evolution. In this instance, *inputs* would consist of external stimuli that generate neuronal activity in the brain. *Thruputs* are analogous to the processing of these inputs in one or more brain regions, depending on neuronal characteristics and circuitry. As in other systems, the degree of *differentiation* (structural and functional specialization) in the brain affects the number of cognitive tools available for processing (i.e., the size and type of neurons and their allocation, interconnectivity, and conduction

velocity). *Feedback* occurs when information is exchanged between brain structures (i.e., the memories established by prior neuronal associations or ensembles) that modulates dissonances, refines interpretation of the stimuli, and formulates an appropriate response. Repeated associations and responses may become imprinted. *Outputs* are the resultant actions taken by the organism that, if adaptive, may evolve into patterns of neurobehavioral specialization (i.e., epigenetic traits). *Equifinality*, in this context, refers to the balancing act pursued by brain structures and functions through time, an ongoing process that continues in the post-Holocene period. See, for example, Weaver and Trinkaus (2005).

Chapter 6. Kinship and Paleolithic Legends

1. See Marx (1965, orig. 1857–58), Bachofen (1861), McLennan (1865), Morgan (1870, 1907, orig. 1877), Lubbock (1873, orig. 1870), Spencer (1896, orig. 1876), and Engels (1968, orig. 1884, 1939, orig. 1888, 1940, orig. 1876).
2. See, for example, Boas (1896, 1911), Kroeber (1909), and Lowie (1920). Ironically, prominent American anthropologists Goldenweiser (1914), Lowie (1920), and Kroeber (1923) registered early objections to the assignment of evolutionary priority to the matriarchate by arguing that Morgan's evolutionary scheme should actually be reversed. Based on their observations of North American Indian cultures, they suggested that matrilineal societies were actually more complex than patrilineal ones, and were therefore the more advanced or evolved form of social organization.
3. See, for example, Radcliffe-Brown and Forde (1950: 72).
4. See Fluehr-Lobban (1979: 345) and Donilova (1971).
5. See also Fluehr-Lobban (1979), Makarius (1977), and Harris's (1968: 637–39) discussion of the Meggers-Opler exchange.
6. If, as the Parkers propose, the bestowal of prestige or "cultural rewards" correlates with the level of risk to life and limb, it could just as easily be argued that the potential dangers faced by Paleolithic women in childbirth may have equaled or exceeded that of Paleolithic men in deer hunting.
7. This passage is also quoted by Knight (2008: 67–68).
8. Borgerhoff-Mulder et al. (2009) made an important contribution by distinguishing three types of transferrable parental investments that enhance fitness: *embodied* wealth (nutritional), *relational* wealth (social networks), and *material* wealth (property). Studies by Smith et al. (2010) and Shennan (2011) separately investigated how these types of intergenerational resource transfer vary among societies of differing levels of socioeconomic complexity. Among their conclusions are that embodied and relational wealth transfers are characteristic of hunter-gatherers, while the accumulation and transfer of divisible material wealth is associated with resource density and sedentism that emerged at the end of the Ice Age. Significantly, parental investments in offspring, even in simple societies, have the potential for wealth stratification and have a measurable impact on reproductive success.

9. See Keverne et al. (1996a, 1999b), Davies, Isles, and Wilkinson (2005), Isles, Davies, and Wilkinson (2006), Keverne and Curley (2008); and Keverne (2017).
10. See Isles et al. (2006) and Keverne and Curley (2008). See also Trivers (1974) for the original parent-offspring conflict theory, which juxtaposes the allocation of resources to offspring with the long-term reproductive goals of the father and the mother.
11. Pennisi (2016) notes that mitochondrial and cell genomes both code for proteins and other molecules that must be compatible, but their disparate replication and mutation rates may create genomic conflicts. One theory is that natural selection spurs mutation rates in the nucleus to keep pace with the mtDNA, thereby restoring compatibility and accelerating the rate of evolutionary change. Others speculate that mtDNA and nuclear incompatibilities are a prelude to population divergence or even speciation.

Chapter 7. Kinship as Social Technology

1. For purposes of discussion in this chapter, the terms "matrilineal" and "patrilineal" shall refer generically to kinship groups formed on the basis of uterine and agnatic linkages, respectively, whether such groups are based on ancestral or contemporary egos. Similarly, since the primary distinctions drawn in these discussions concern the consequences of localizing female uterines versus male agnates, and due to their frequently interchangeable usage in the literature, the term matrilocal residence is equated here with uxorilocal residence and the term patrilocal residence with virilocal residence.
2. See Lippert (1931: 237), Linton (1936: 168–69), Murdock (1949: 204–06), Driver and Massey (1957: 425), Steward (1955: 125), and Steward and Faron (1959: 384).
3. Divale (1974: 14) reports that, based on a sample of 1,153 societies coded in the *Ethnographic Atlas*, patrilocal/virilocal societies predominate over matrilineal/uxorilocal societies by a ratio of 70 percent to 15 percent. As noted by Martin (1974), however, this sample does not include matrilineal foragers in the South American Gran Chaco region and underrepresents the frequency of uterine social organization in this and perhaps other regions in precontact times.
4. See Trigger (1976, 1978), Ember (1973), and Divale (1977).
5. See Steward and Faron (1959) and Martin (1969). South American foragers have a very high incidence of uterine organization but are poorly represented in the *Ethnographic Atlas*. Matrilineal descent in the Gran Chaco region occurred among pedestrian hunter-gatherer groups such as the Zamuco, Chamacoco, Mataco, Choroti, Ashluslay, Maca, Lengua-Enimago, Lule-Vilelans, Tupians (Tapiete), Lengua, Kaskiha, Siriono, Guayaki, Yaruro, and Warrau; and among equestrians such as the Mocovi, Mbaya, Abipon, and Payagua. Matrilocal residence obtained among all except the matri-patrilocal Abipon and ambilocal Mocovi, both of which had endog-

amous multilineage communities (data for the Payagua and Tupians are unavailable). Only the Siriono, Guayaki, Yaruro, and Warrau undertake sporadic gardening and appear to be devolved horticulturalists. Patrilocality is specifically reported only for the Puelche and Tehuelche, and even in these cases there is some question regarding the aboriginality of this organization. For example, patrilocality was reported for the Puelche and Tehuelche in 1830, about 160 years after their acquisition of the horse.

6. Watson-Franke went on to suggest that the matrilineal male model would perhaps better serve modern Western society in terms of both parenting and the treatment of women. This led to an exchange of comments with Bolyanatz (1995a, 1995b), who criticized Watson-Franke's conclusions as polemical (read feminist). Based on his fieldwork in New Ireland, he questioned the notion that matriliny produces a distinct set of values and a favorable cognitive focus on women by noting that Sursurunga men occasionally exhibit socially dominant or aggressive behaviors toward their wives and sisters. Bolyanatz also questioned the existence of a matrilineal puzzle, but on the basis that matrilineal males regularly work out conflicts between their descent and affinal roles among themselves. In her response, Watson-Franke (1995) reiterated that matrilineal male roles are highly variable—that her model never claimed that matrilineal men made better husbands and fathers, or that they never beat their wives. One size does not fit all, either geographically or historically. Significantly, as Watson-Frank points out, the New Ireland populations among whom Bolyanatz worked changed their postmarital residence practices to virilocality as a direct result of colonial policies (i.e., a pattern corresponding to Schlegel's Husband-Dominant type).

7. See, for example, Holden, Sear, and Mace (2003), Fortunato and Archetti (2010), and Fortunato (2012).

Bibliography

Aberle, David F. 1961. "Matrilineal Descent in Cross-Cultural Perspective." In *Matrilineal Kinship*, ed. David M. Schneider and Kathleen Gough, 655–727. Berkeley and Los Angeles: University of California Press.

Aiello, L., and R. Dunbar. 1993. "Neocortex Size, Group Size and the Evolution of Language." *Current Anthropology* 34: 184–93.

Aiello, Leslie C., and Cathy Key. 2002. "Energetic Consequences of Being a *Homo erectus* Female." *American Journal of Human Biology* 14: 551–65.

Aiello, Leslie C., and Peter Wheeler. 1995. "The Expensive-Tissue Hypothesis: The Brain and the Digestive Systems in Human and Primate Evolution." *Current Anthropology* 36(2): 199–221.

Alberts, S. C. 1999. "Paternal Kin Discrimination in Wild Baboons." *Proceedings of the Royal Society B* 266: 1501–6.

Alexander, R.D. 1974. "The Evolution of Social Behavior." *Annual Review of Ecology and Semantics* 5: 325–83.

Alexander, Richard D., and Katherine Noonan. 1979. "Concealment of Ovulation, Parental Care, and Human Social Evolution." In *Evolutionary Biology and Human Social Behavior*, ed. N. Chagnon and W. Irons, 436–53. North Scituate, Mass: Duxbury Press.

Allen, Benjamin, Martin A. Nowak, and Edward O. Wilson. 2013. "Limitations of Inclusive Fitness." *Proceedings of the National Academies of Sciences USA* 110(50): 20135–39.

Allen, Nicholas J. 2008. "Tetradic Theory on Human Kinship Systems." In *Early Human Kinship: From Sex to Social Reproduction*, ed. Nicholas J. Allen et al., 96–112. Oxford: Blackwell.

Allen, Nicholas J., Hilary Callan, Robin Dunbar, and Wendy James, eds. 2008. *Early Human Kinship: From Sex to Social Reproduction*. Oxford: Blackwell.

Allman, J. M., et al. 2001. "The Anterior Cingulate Cortex. The Evolution of an Interface Between Emotion and Cognition." *Annals of the New York Academy of Sciences* 935: 107–17.

Allman, J., A. Hakeem, and K. Watson. 2002. "Two Phylogenetic Specializations in the Human Brain." *The Neuroscientist* 8(4): 335–46.

Alperzon-Afil, Nira. 2008. "Continual Fire-Making by Hominins at Gesher Benot Ya'agov." *Quarterly Science Reviews* 27(17–18): 1733–39.

Amici, Federica, Filippo Aureli, and Josep Call. 2008. "Fission-Fusion Dynamics, Behavioral Flexibility, and Inhibitory Control in Primates." *Current Biology* 18: 1415–19.

Anton, Susan C. 2003. "Natural History of *Homo erectus*." *American Journal of Physical Anthropology* 122, Supplement 37: 126–70.

Anton, Susan C. 2012. "Early *Homo*: Who, Where and When." *Current Anthropology* 53(S6): S278–98.

Anton, Susan C., W. R. Leonard, and M. L. Robertson. 2002. "An Ecomorphological Model of the Initial Hominid Dispersal from Africa." *Journal of Human Evolution* 43: 773–85.

Anton, Susan C., Richard Potts, and Leslie C. Aiello. 2014. "Evolution of Early *Homo*: An Integrated Biological Perspective." *Science* 345(6192): 1236828.
Anton, Susan C., and J. Josh Snodgrass. 2012. "Origins and Evolution of Genus *Homo*: New Perspectives." *Current Anthropology* 53(S6): S479–96.
Appenzeller, Tim. 2018. "Europe's First Artists Were Neandertals." *Science* 359(6378): 852–53.
Ardrey, Robert. 1961. *African Genesis: A Personal Investigation into the Animal Origins and Nature of Man*. New York: Antheneum.
———. 1966. *The Territorial Imperative*. New York: Antheneum Press.
———. 1976. *The Hunting Hypothesis: A Personal Conclusion Concerning the Evolutionary Nature of Man*. New York: Antheneum.
Arnold, Carrie. 2017. "Choosy Eggs May Pick Sperm for Their Genes, Defying Mendel's Law." *Quanta Magazine* (November 15). https://www.quantamagazine.org/choosy-eggs-may-pick-sperm-for-their-genes-defying-mendels-law-20171115/.
Arsuaga, J. L., et al. 2014. "Neanderthal Roots: Cranial and Chronological Evidence from Sima De Los Huesos." *Science* 344 (6190): 1358–63.
Ashley, Gail M. 2007. "Orbital Rhythms, Monsoons, and Playa Lake Response, Olduvai Basin, Equatorial East Africa (ca. 1.85-1.74 Ma)." *Geology* 35(12): 1091–94.
Bachofen, Johann. 1861. *Das Mutterrecht*. Basil: Benno Schwabe.
Bae, Christopher J., Katerina Douka, and Michael D. Petraglia. 2017. "On the Origin of Modern Humans: Asian Perspectives." *Science* 358: 1269–75.
Barlow, George W., and James Silverberg, eds. 1980. *Sociobiology: Beyond Nature/Nurture? Reports, Definitions and Debate*. Boulder: Western Press.
Barnard, Alan. 2008. "The Co-Evolution of Language and Kinship." In *Early Human Kinship: From Sex to Social Reproduction*, ed. Nicholas J. Allen et al., 232–43. Oxford: Blackwell.
———. 2009. "Social Origins: Sharing, Exchange, Kinship." In *The Cradle of Language*, ed. Rudolf Botha and Chris Knight, 29–35. Oxford: Oxford University Press.
———. 2011. *Social Anthropology and Human Origins*. Cambridge: Cambridge University Press.
Barton, Robert A., and Paul H. Harvey. 2000. "Mosaic Evolution of Brain Structure in Mammals." *Nature* 405: 1055–58.
Beaumont, Peter B. 2011. "The Edge: More on Fire-Making by About 1.7 Million Years Ago at Wonderwerk Cave in South Africa." *Current Anthropology* 52(4): 585–95.
Begler, Elsie B. 1978. "Sex, Status, and Authority in Egalitarian Society." *American Anthropologist* 80(3): 571–88.
Belmaker, M., E. Tchernov, S. Condemi, and O. Bar-Yosef. 2002. "New Evidence for Hominid Presence in the Lower Pleistocene of the Southern Levant." *Journal of Human Evolution* 43: 43–56.
Berger, Lee, et al. 2015. "*Homo naledi*: A New Species of the Genus *Homo* from the Dinaledi Chamber, South Africa." doi.10.7554/eLife.09560.001.
Berger, Lee, and John Hawkes. 2017. *Almost Human*. Washington, DC: National Geographic Partners.
Binford, L.R. 1981. *Bones: Ancient Man and Modern Myths*. New York: Academic Press.

———. 1990. "Mobility, Housing, and Environment: A Comparative Study." *Journal of Anthropological Research* 46: 119–52.
Binford, Sally. 1968. "Early Upper Pleistocene Adaptations in the Levant." *American Anthropologist* 70: 707–17.
Boas, Franz. 1896. "The Limitations of the Comparative Method of Anthropology." In *Race, Language and Culture*, 271–304. New York: Macmillan.
———. 1911. *The Mind of Primitive Man*. New York: Macmillan.
Boaz, Noel. 1993. *Quarry: Closing in on the Missing Link*. New York: The Free Press.
———. 1997. *Eco Homo: How the Human Being Emerged from the Cataclysmic History of the Earth*. New York: Basic Books.
Boaz, Noel T., ed. 1990. *Evolution of Environments and Hominidae in the African Western Rift Valley*. Martinsville, Va: Virginia Museum of Natural History Memoir Number 1.
Boaz, Noel, and Russell L. Ciochon. 2004. *Dragon Bone Hill: An Ice-Age Saga of Homo Erectus*. New York: Oxford University Press.
Bolyanatz, Alexander H. 1995a. "Matriliny as Revisionist Anthropology." *Anthropos* 90: 169–80.
———. 1995b. "Second Reply to Watson-Franke." *Anthropos* 90: 585–86.
Borgerhoff-Mulder, Monique, et al. 2009. "The Intergenerational Transmission of Wealth and the Dynamics of Inequality in Pre-Modern Societies." *Science* 326: 682–88.
Boserup, Ester. 1970. *Women's Role in Economic Development*. London: G. Allen and Unwin.
Bossdorf, O., C. L. Richards, and M. Pigliucci. 2008. "Epigenetics for Ecologists." *Ecology Letters* 11: 106–15.
Braun, David R., et al. 2010. "Early Hominin Diet Included Diverse Terrestrial and Aquatic Animals 1.95 Ma in East Turkana, Kenya." *Proceedings of the National Academy of Sciences* 107(22): 10002–07.
Bribiescas, Richard G., Peter T. Ellison, and Peter B. Gray. 2012. "Male Life History, Reproductive Effort, and the Evolution of the Genus *Homo*." *Current Anthropology* 53(6): S424–35.
Briffault, R. 1963 (orig. 1927). *The Mothers*. Abridged with an introduction by G. Rattray Taylor. New York: Universal Library.
Broadhurst, C. Leigh, Stephen C. Cunnane, and Michael A. Crawford. 1998. "Rift Valley Lake Fish and Shellfish Provided Brain-Specific Nutrition for Early Homo." *British Journal of Nutrition* 79: 3–21.
Broadhurst, C. Leigh, et al. 2002. "Brain-Specific Lipids from Marine, Lacustrine, or Terrestrial Food Resources: Potential Impact on Early African *Homo sapiens*." *Comparative Biochemistry and Physiology Part B: Biochemistry and Molecular Biology* 131(4): 653–73.
Broadhurst, C. Leigh, Michael Crawford, and Stephen Monroe. 2011. "Littoral Man and Waterside Woman: The Crucial Role of Marine and Lacustrine Foods and Environmental Resources in the Origin, Migration and Dominance of *Homo Sapiens*." In *Was Man More Aquatic in the Past? Fifty Years After Alister Hardy Waterside Hypotheses of Human Evolution*, ed. Mario Vaneechoutee, Algis Kuliukas, and Marc Verhaegen, 16–35. New York: Bentham Books.

Brodman, K. 1912. "Neue Ergebnisse Uber die Vergleichende Histologische Lokalisation der Grosshirnrinde mit Besonderer Berucksichtigung des Stirnhirns." *Anatomischen Anzeiger* 41: 157–216.
Broom, Robert. 1933. *The Coming of Man: Was It Accident or Design?* London: Witherby, HF and Witherby, G.
Brown, Judith K. 1970. "Economic Organization and the Position of Women Among the Iroquois." *Ethnohistory* 17: 151–67.
Bruggemann, J. Heinrich, et al. 2004. "Stratigraphy, Palaeoenvironments and Model for the Deposition of the Abdur Reef Limestone: Context for an Important Site from the Last Interglacial on the Red Sea Coast of Eritrea." *Palaeogeography, Palaeoclimatology, Palaeoecology* 203(3–4): 179–352.
Buskirk, J. Van, and U. K. Steiner. 2009. "The Fitness Costs of Developmental Canalization and Plasticity." *Journal of Evolutionary Biology* 22: 852–60.
Byrne, R. W., and A. Whitten, eds. 1988. *Machiavellian Intelligence*. Oxford, UK: Oxford University Press.
Cachel, Susan. 1997. "Dietary Shifts and the European Upper Paleolithic Transition." *Current Anthropology* 38(4): 579–603.
Callan, Hilary. 2008. "Reaching Across the Gap." In *Early Human Kinship: From Sex to Social Reproduction*, ed. Nicholas J. Allen et al., 247–58. Oxford: Blackwell.
Calvin, William H., and Derek Bickerton. 2000. *Lingua ex machina: Reconciling Darwin and Chomsky with the Human Brain*. Cambridge, Mass.: MIT Press.
Carbonell, E., et al. 1995. "Lower Pleistocene Hominids and Artifacts from Atapuerca-TD6 (Spain)." *Science* 269(5225): 826–30.
Carillo-Reid, Luis, et al. 2016. "Imprinting and Recalling Cortical Ensembles." *Science* 353 (6300): 691–94.
Carlson, Bryce A., and John D. Kingston. 2007. "Docosahexaenoic Acid Biosynthesis and Dietary Contingency: Encephalization Without Aquatic Constraint." *American Journal of Human Biology* 19(4): 585–88.
Chance, M.R.A. 1961. "The Nature and Special Features of the Instinctive Social Bonds of Primates." In *The Social Life of Early Man*, ed. Sherwood Washburn, 17–33. Chicago: Aldine Publishing Company.
Chapais, Bernard. 2008. *Primeval Kinship: How Pair-Bonding Gave Birth to Human Society*. Cambridge, Mass.: Harvard University Press.
———. 2011. "The Deep Social Structure of Humankind." *Science* 331: 1276–77.
———. 2014. "Complex Kinship Patterns as Evolutionary Constructions, and the Origins of Sociocultural Universals." *Current Anthropology* 55(6): 751–81.
Chapman, Colin A., and Lauren J. Chapman. 2000. "Determinants of Group Size in Primates: The Importance of Travel Costs." In *On the Move: How and Why Animals Travel in Groups*, ed. S. Boinski and P. A. Garber, 24–41. Chicago: Chicago University Press.
Chen, Feng-Chi, and Wen-Hsiung Li. 2001. "Genomic Differences Between Humans and Other Hominoids and the Effective Population Size of the Common Ancestor of Humans and Chimpanzees." *American Journal of Human Genetics* 68: 444–56.
Chen, Ingfei. 2009. "Brain Cells for Socializing." *Smithsonian Magazine*. June Issue. www.smithsonianmag.com/science-nature/brain-cells-for-socializing-133855450/.

Claassen, C. 1991. "Normative Thinking and Shell-Bearing Sites." *Archaeological Method and Theory* 3: 249–98.
Clancy, Kelly. 2017. "Survival of the Friendliest." *Nautilus*: Issue 046, Chapter 4. http://nautil.us/issue/46/balance/survival-of-the-friendliest.
Clark, J. D. 1980. "Early Human Occupation of African Savanna Environments." In *Human Ecology in Savanna Environments*, ed. D. R. Harris, 41–71. London: Academic Press.
Clutton-Brock, T.H., and P.H. Harvey. 1976. "Evolutionary Rules and Primate Societies." In *Growing Points in Ethology*, ed. P.P.G. Bateson and R.A. Hinde, 195–237. Cambridge: Cambridge University Press.
Colonese, A. C. et al. 2011. "Marine Mollusc Exploitation in Mediterranean Prehistory: An Overview." *Quaternary International* 239(1–2): 86–103.
Conard, Nicholas J., et al. 2015. "Excavations at Schöningen and Paradigm Shifts in Human Evolution." *Journal of Human Evolution* 89: 1–17.
Coolidge, Frederick L., and Thomas Wynn. 2009. *The Rise of Homo Sapiens: The Evolution of Modern Thinking*. Oxford: Wiley-Blackwell.
Copeland, Sandi R., et al. 2011. "Strontium Isotope Evidence for Landscape Use by Early Hominins." *Nature* 474: 76–78.
Copes, Lynn E., and William H. Kimbel. 2016. "Cranial Vault Thickness in Primates: *Homo erectus* Does Not Have Uniquely Thick Vault Bones." *Journal of Human Evolution* 90: 120–34.
Crawford, M. A., et al. 1999. "Evidence for the Unique Function of Docosahexaenoic Acid During the Evolution of the Modern Hominid Brain." *Lipids* 34: 539–47.
Cunnane, Stephen C., et al. 2007. "Docosahexaenoic Acid and Shore-Based Diets in Hominin Encephalization: A Rebuttal." *American Journal of Human Biology* 19(4): 578–81.
Daly, M., and M. Wilson. 1978. *Sex, Evolution and Behavior*. North Scituate: Duxbury Press.
Dart, Raymond. 1925. "*Australopithecus africanus*: The Man-Ape of South Africa." *Nature* (London) 115: 195–99.
———. 1953. "The Predatory Transition from Ape to Man." *International Anthropological and Linguistic Review* 1(4): 201–17.
Darwin, Charles. 1859. *The Origin of Species by Natural Selection*. London: Murray.
———. 1871. *The Descent of Man and Selection in Relation to Sex*. London: Murray.
Davies, William, Anthony R. Isles, and Lawrence S. Wilkinson. 2005. "Imprinted Gene Expression in the Brain." *Neuroscience and Biobehavioral Reviews* 29(3): 421–30.
Dawkins, Richard. 1976. *The Selfish Gene*. New York: Oxford University Press.
———. 1982. *The Extended Phenotype*. New York: Oxford University Press.
Debetz, G.F. 1961. "The Social Life of Early Paleolithic Man as Seen Through the Work of Soviet Anthropologists." In *The Social Life of Early Man*, ed. Sherwood Washburn, 137–49. Chicago: Aldine Publishing Company.
de Menocal, Peter B. 2004. "African Climate Change and Faunal Evolution During the Pliocene-Pleistocene." *Earth and Science Letters* 220: 3–24.
de Ruiter, Jan, Gavin Weston, and Stephen M. Lyon. 2011. "Dunbar's Number: Group Size and Brain Physiology in Humans Reexamined." *American Anthropologist* 113(4): 557–68.

de Vladar, Harold P., and Eors Szathmary. 2017. "Beyond Hamilton's Rule." *Science* 356(6337): 485–86.
de Waal, Frans. 2005. *Our Inner Ape*. New York: Riverhead Books.
———. 2007 (orig. 1982). *Chimpanzee Politics: Power and Sex Among Apes*. Baltimore: Johns Hopkins University Press.
———. 2013. *The Bonobo and the Atheist: In Search of Humanism Among the Primates*. New York: W. W. Norton.
———. 2016. *Are We Smart Enough to Know How Smart Animals Are?* New York: W.W. Norton & Company.
de Winter, Willem, and Charles E. Oxnard. 2000. "Evolutionary Radiations and Convergences in the Structural Organization of Mammalian Brains." *Nature* 409: 710–14.
Diamond, Jared. 1997. *Guns, Germs and Steel: A Short History of Everybody for the Last 13,000 Years*. London: Jonathan Cape.
Dickeman, Mildred. 1979. "Comment on van den Berghe's and Barash's Sociobiology." *American Anthropologist* 81: 351–57.
Dickens, Thomas E., and Qazi Rahman. 2012. "The Extended Evolutionary Synthesis and the Role of Soft Inheritance in Evolution." *Proceedings of the Royal Society B* 279 (1740): 1–9.
Divale, William T. 1974. "Migration, External Warfare, and Matrilocal Residence." *Behavior Science Research* 9: 75–133.
———. 1977. "Living Floor and Marital Residence: A Replication." *Behavior Science Research* 12: 109–15.
———. 1984. *Matrilocal Residence in Pre-Literate Society*. Ann Arbor, Michigan: UMI Research Press.
Divale, William T., and Marvin Harris. 1976. "Population, Warfare, and the Male Supremacist Complex." *American Anthropologist* 78(3): 521–38.
Dobzhansky, T. 1937. *Genetics and the Origin of Species*. New York: Columbia University Press.
Dominguez-Rodrigo, M. 2014. "Is the 'Savanna Hypothesis' a Dead Concept for Explaining the Emergence of the Earliest Hominins?" *Current Anthropology* 55(1): 59–81.
Donilova, L. V. 1971. "Controversial Problems on the Theory of Pre-Capitalist Society." *Soviet Anthropology and Archaeology* 9: 269–328.
Driver, Harold. 1961. *Indians of North America*. Chicago: University of Chicago Press.
Driver, Harold, and William Massey. 1957. "Comparative Studies of North American Indians." *Transactions of the American Philosophical Society* 47: 165–460.
Dublin, Holly T. 1983. "Cooperative and Reproductive Competition Among Female African Elephants." In *Social Behavior of Female Vertebrates*, ed. Samuel K. Wasser, 291–313. New York: Academic Press.
Dunbar, R. I. M. 1988. "The Social Brain Hypothesis." *Evolutionary Anthropology* 6: 178–90.
———. 1992. "Neocortex Size as a Constraint on Group Size in Primates." *Journal of Human Evolution* 22: 469–93.
———. 1993. "Coevolution of Neocortex Size, Group Size and Language in Humans." *Behavioral and Brain Sciences* 16: 681–735.

———. 1995. "Neocortex Size and Group Size in Primates: A Test of the Hypothesis." *Journal of Human Evolution* 28: 287–96.
———. 2003. "The Social Brain: Mind, Language, and Society in Evolutionary Perspective." *Annual Reviews of Anthropology* 32: 163–81.
———. 2008. "Kinship in Biological Perspective." In *Early Human Kinship: From Sex to Social Reproduction*, ed. Nicholas J. Allen et al., 131–50. Oxford: Blackwell.
———. 2009. "Why Only Humans Have Language." In *The Prehistory of Language*, ed. R. Botha and C. Knight, 12–35. Oxford: Oxford University Press.
Ellis, Richard. 2011. "Aquagenesis: Alister Hardy, Elaine Morgan and the Aquatic Ape Hypothesis." In *Was Man More Aquatic in the Past? Fifty Years After Alister Hardy Waterside Hypotheses of Human Evolution*, ed. Mario Vaneechoutee, Algis Kuliukas, and Marc Verhaegen, 190–98. New York: Bentham Books.
Ember, Melvin. 1973. "An Archaeological Indicator of Matrilocal Versus Patrilocal Residence." *American Antiquity* 38: 177–82.
Ember, Melvin, and Carol R. Ember. 1971. "The Conditions Favoring Matrilocal Versus Patrilocal Residence." *American Anthropologist* 73: 571–94.
Emlen, Stephen T., and Lewis W. Oring. 1977. "Ecology, Sexual Selection, and the Evolution of Mating Systems." *Science* 197 (4300): 215–33.
Engels, Frederick. 1940 (orig. 1876). *Dialectics of Nature*. New York: International Publishers.
———. 1968 (orig. 1884). *Origin of the Family, Private Property and the State*. Moscow: Progress Publishers.
———. 1939 (orig. 1888). *Eugene Duhring's Revolution in Science (Anti-Duhring)*. New York: International Publishers.
Erlandson, Jon M. 2001. "The Archaeology of Aquatic Adaptations: Paradigms for a New Millennium." *Journal of Archaeological Research* 9(4): 287–350.
———. 2010. "Food for Thought: The Role of Coastlines and Aquatic Resources in Human Evolution." In *Human Brain Evolution*, ed. Stephen C. Cunnane and Kathlyn M. Stewart, 125–36. New York: John Wiley and Sons, Inc.
Ferring, Reid, et al. 2011. "Earliest Human Occupations at Dmanisi (Georgian Caucasus)." *Proceedings of the National Academy of Sciences* 108(26): 10432–36.
Finlayson, Clive. 2004. *Neanderthals and Modern Humans: An Ecological and Evolutionary Perspective*. New York: Cambridge University Press.
———. 2009. *The Humans Who Went Extinct: Why Neanderthals Died Out and We Survived*. New York: Oxford University Press.
———. 2014. *The Improbable Primate: How Water Shaped Human Evolution*. New York: Oxford University Press.
Finnegan, Morna. 2017. "Who Sees the Elephant? Sexual Egalitarianism in Social Anthropology's Room." In *Human Origins: Contributions from Social Anthropology*, ed. Camilla Power, Morna Finnegan, and Hilary Callan, 130–52. New York: Berghahn.
Fix, Alan G. 1976. "Comment on John Hartung, et al., 'On Natural Selection and the Inheritance of Wealth'." *Current Anthropology* 17(4): 616–17.
Flatt, Thomas. 2005. "The Evolutionary Genetics of Canalization." *Quarterly Review of Biology* 80(3): 287–316.

Fluehr-Lobban, Carolyn. 1979. "A Marxist Reappraisal of the Matriarchate." *Current Anthropology* 20(2): 341–59.

Foley, R. A., and P. C. Lee. 1989. "Finite Social Space, Evolutionary Pathways, and Reconstructing Hominid Behavior." *Science* 243: 901–6.

Foley, R. A., and P. C. Lee. 1991. "Ecology and Energetics of Encephalization in Hominid Evolution." *Philosophical Transactions of the Royal Society of London B* 334: 223–32.

Fortes, Meyer. 1969. *Kinship and the Social Order.* London: Routledge and Kegan Paul.

Fortunato, Laura. 2012. "The Evolution of Matrilineal Kinship Organization." *Proceedings of the Royal Society B* 279: 4939–45.

Fortunato, Laura, and M. Archetti. 2010. "Evolution of Monogamous Marriage by Maximization of Inclusive Fitness." *Journal of Evolutionary Biology* 23: 149–56.

Foster, Kevin R., and Francis L. W. Ratnieks. 2005. "A New Eusocial Vertebrate?" *Trends in Ecology and Evolution* 20: 363–64.

Fox, Robin. 1967. *Kinship and Marriage: An Anthropological Perspective.* Baltimore: Penguin Books.

———. 1972. "Alliance and Constraint: Sexual Selection in the Evolution of Human Kinship Systems." In *Sexual Selection and the Descent of Man 1871-1971*, ed. B.G. Campbell, 282–331. Chicago: Aldine Publishing Company.

———. 1991. "Comment" on Rodseth et al., "The Human Community as a Primate Society." *Current Anthropology* 32(3): 242–43.

Gabunia, Leo, et al. 2000. "Earliest Pleistocene Hominid Cranial Remains from Dmanisi, Republic of Georgia: Taxonomy, Geological Setting, and Age." *Science* 288(5468): 1019–25.

Gamble, Clive. 1998. "Paleolithic Society and the Release from Proximity: A Network Approach to Intimate Relations." *World Archaeology* 29(3): 426–49.

———. 2008. "Kinship and Material Culture: Archaeological Implications of the Human Global Diaspora." In *Early Human Kinship: From Sex to Social Reproduction*, ed. Nicholas J. Allen et al., 27–40. Oxford: Blackwell.

Ghiglieri, Michael P. 1987. "Sociobiology of the Great Apes and the Hominid Ancestor." *Journal of Human Evolution* 16: 319–57.

Gibbons, Ann. 2007. "Food for Thought." *Science* 316(5831): 1558–60.

———. 2013. "How a Fickle Climate Made Us Human." *Science* 341(6145): 474–79.

———. 2017. "Oldest Members of Our Species Discovered in Morocco." *Science* 356(6342): 993–94.

Gibert, Luis, et al. 2016. "Chronology for the Cueva Victoria Fossil Site (SE Spain): Evidence for Early Pleistocene Afro-Iberian Dispersals." *Journal of Human Evolution* 90: 183–97.

Gintis, Herbert. 2011. "Gene-Culture Coevolution and the Nature of Human Sociality." *Philosophical Transactions of the Royal Society B* 366: 878–88.

Goldenweiser, A. A. 1914. "The Social Organization of the Indians of North America." *Journal of American Folk-Lore* 27: 436.

Goodenough, Ward. 1963. "Book Review of Matrilineal Kinship (David Schneider and Kathleen Gough, eds., Berkeley and Los Angeles: University of California Press, 1971)," *American Anthropologist* 65(4): 923–28.

Gordon, Adam D., Steig F. Johnson, and Edward E. Louis Jr. 2013."Females Are the Ecological Sex: Sex-Specific Body Mass Ecogeography in Wild Sifaka Populations (*Propithecus* spp.)." *American Journal of Physical Anthropology* 151(1): 77–87.
Goren-Inbar, N. 1986. "A Figurine from the Acheulean Site of Berekhat Rom." *Mitakufat Haeven* 19: 7–12.
Gough, Kathleen 1961a. "Variation in Residence." In *Matrilineal Kinship*, ed. David Schneider and Kathleen Gough, 545–76. Berkeley and Los Angeles: University of California Press.
———. 1961b. "The Modern Disintegration of Matrilineal Descent Groups." In *Matrilineal Kinship*, ed. David Schneider and Kathleen Gough, 631–52. Berkeley and Los Angeles: University of California Press.
Gowlett, John A. J. 2008. "Deep Roots of Kin: Developing the Evolutionary Perspective from Prehistory." In *Early Human Kinship: From Sex to Social Reproduction*, ed. Nicholas J. Allen et al., 41–57. Oxford: Blackwell.
Gowlett, John A. J., and Robin Dunbar. 2008. "A Brief Overview of Human Evolution." In *Early Human Kinship: From Sex to Social Reproduction*, ed. Nicholas J. Allen et al., 21–24. Oxford: Blackwell.
Gowlett, John, Clive Gamble, and Robin Dunbar. 2012. "Human Evolution and the Archaeology of the Social Brain." *Current Anthropology* 53(6): 693–722.
Greuter, Cyril C., Bernard Chapais, and Dietmar Zinner. 2012. "Evolution of Multilevel Social Systems in Nonhuman Primates and Humans." *International Journal of Primatology* 33: 1002–37.
Gusinde, Martin. 1931. *Die Selk'nam*. Modling bei Wien.
———. 1961. *The Yamana (Die Yamana)*, trans. Frieda Schutze. New Haven: Human Relations Area Files.
Haig, David. 1997. "Parental Antagonism, Relatedness Asymmetries, and Genomic Imprinting." *Proceedings of the Royal Society of London B* 264: 1657–62.
———. 2000. "Genomic Imprinting, Sex-Biased Dispersal, and Social Behavior." *Annals of the New York Academy of Sciences* 907(1): 149–63.
Hamilton, Marcus J., et al. 2007a. "Nonlinear Scaling of Space Use in Human Hunter-Gatherers." *Proceedings of the National Academy of Sciences* 104(11): 4765–69.
Hamilton, Marcus J., et al. 2007b. "The Complex Structure of Hunter-Gatherer Social Networks." *Proceedings of the Royal Society B* 274: 2195–2202.
Hamilton, W. D. 1963. "The Evolution of Social Behavior." *The American Naturalist* 97(896): 354–56.
———. 1964. "The Genetical Evolution of Social Behavior (I and II)." *Journal of Theoretical Biology* 7:1–16, 17–52.
Harari, Yuval Noah. 2011. *Sapiens: A Brief History of Mankind*. London: Vintage.
Hardy, Alister. 1960. "Was Man More Aquatic in the Past?" *New Scientist* 7: 642–45.
Harris, Marvin. 1968. *The Rise of Anthropological Theory*. New York: Thomas Y. Crowell Company.
———. 1989. *Our Kind: Who We Are, Where We Came From, Where We Are Going*. New York: Harper & Row.

Hart, John P. 1999. "Maize Agriculture Evolution in the Eastern Woodlands of North America: A Darwinian Perspective." *Journal of Archaeological Method and Theory* 6: 137–80.

———. 2000. "New Dates from Classic New York Sites: Just How Old Are Those Longhouses?" *Northeast Anthropology* 60: 1–22.

———. 2001. "Maize, Matrilocality, Migration, and Northern Iroquoian Evolution." *Journal of Archaeological Method and Theory* 8(2): 151–82.

Hartung, John. 1985. "Matrilineal Inheritance: New Theory and Analysis." *Behavioral and Brain Sciences* 8(4): 661–70.

Hartung, John, et al. 1976. "On Natural Selection and the Inheritance of Wealth." *Current Anthropology* 17(4): 607–22.

Hawkes, K. 1990. "Why Do Men Hunt? Some Benefits for Risky Strategies." In *Risk and Uncertainty in Tribal and Peasant Economies*, ed. E. Cashden, 145–66. Boulder, CO: Westview Press.

———. 1993. "Why Hunter-Gatherers Work: An Ancient Version of the Problem of Public Goods." *Current Anthropology* 34: 341–61.

Hawkes, K., J. F. O'Connell, and N.G. Blurton Jones. 1991. "Hunting Income Patterns Among the Hadza: Big Game, Common Goods, Foraging Goals, and the Evolution of the Human Diet." *Philosophical Transactions of the Royal Society of London B* 334: 243–51.

Hawkes, K., J. F. O'Connell, and N.G. Blurton Jones. 1997. "Hadza Women's Time Allocation, Offspring Provisioning, and the Evolution of Post-Menopausal Lifespans." *Current Anthropology* 38: 551–78.

Hawkes, K., et al. 1998. "Grandmothering, Menopause, and the Evolution of Human Life Histories." *Proceedings of the National Academy of Sciences* 95(3): 1336–39.

Hawkes, K., and R. Bliege Bird. 2002. "Showing Off, Handicap Signaling, and the Evolution of Men's Work." *Evolutionary Anthropology* 11: 58–67.

Hawkes, K., and R. Bliege Bird. 2004. "The Grandmother Effect." *Nature* 428: 128–29.

Hawkes, K., and N. Blurton Jones. 2005. "Human Age Structures, Paleodemography, and the Grandmother Hypothesis." In *Grandmotherhood: The Evolutionary Significance of the Second Half of Female Life*, ed. E. Voland, A. Chasiotis, and W. Schiefenhovel, 118–43. New Brunswick, New Jersey: Rutgers University Press.

Hebb, D. O. 1949. *The Organization of Behavior*. New York: Wiley & Sons.

Henry, Amanda G., Alison S. Brooks, and Dolores R. Piperno. 2011. "Microfossils in Calculus Demonstrate Consumption of Plants and Cooked Foods in Neanderthal Diets (Shanidar III, Iraq; Spy I and II, Belgium)." *Proceedings of the National Academy of Sciences* 108(2): 486–91.

Henry, Amanda G., Alison S. Brooks, and Dolores R. Piperno. 2014. "Plant Foods and the Dietary Ecology of Neanderthals and Early Modern Humans." *Journal of Human Evolution* 69: 44–54.

Herculano-Houzel, Suzana. 2009. "The Human Brain in Numbers: A Linearly Scaled-Up Primate Brain." *Human Neuroscience* 3: 1–11.

Herculano-Houzel, Suzana, Paul R. Manger, and Jon H. Kaas. 2014. "Brain Scaling in Mammalian Evolution as a Consequence of Concerted and Mosaic

Changes in Numbers of Neurons and Average Neuronal Cell Size." *Frontiers in Neuroanatomy* 8: 1–28.

Hill, Kim R., et al. 2011. "Co-Residence Patterns in Hunter-Gatherer Societies Show Unique Human Social Structure." *Science* 331: 1286–89.

Hoberg, E., N. L. Alkire, A. de Queiroz, and A. Jones. 2001. "Out of Africa: Origins of the *Taenia* Tapeworms in Humans." *Proceedings of the Royal Society of Biological Sciences* 268(1469): 781–87.

Hoffmann, D. L., et al. 2018. "U-Th Dating of Carbonate Crusts Reveals Neandertal Origin of Iberian Cave Art." *Science* 359(6378): 912–15.

Holden, Claire Janaki, Rebecca Sear, and Ruth Mace. 2003. "Matriliny as Daughter-Based Investment." *Evolution and Human Behavior* 24: 99–112.

Holliday, Trenton W. 1998. "The Ecological Context of Trapping Among Recent Hunter-Gatherers: Implications for Subsistence in Terminal Pleistocene Europe." *Current Anthropology* 19(5): 711–19.

Holloway, Ralph L. 1972a. "New Australopithecine Endocast, SK1585, from Swartkrans, South Africa." *American Journal of Physical Anthropology* 37: 173–85.

———.1972b. "Australopithecine Endocasts, Brain Evolution in the Hominoidea, and a Model of Hominid Evolution." In *The Functional and Evolutionary Biology of Primates,* ed. Russell H. Tuttle, 185–203. Chicago: Aldine.

———. 1976. "Paleoneurological Evidence for Language Origins." In *Origins and Evolution of Language and Speech,* ed. S. R. Harnad, H. D. Steklis, and J. Lancaster, 330–48. Annals of the New York Academy of Sciences, Vol. 280. New York: New York Academy of Sciences.

Hrdy, Sarah B. 1981. *The Woman Who Never Evolved.* Cambridge, Mass: Harvard University Press.

———. 1984. "Female Reproductive Strategies." In *Female Primates: Studies by Women Primatologists,* ed. Meredith F. Small, 103–9. New York: Alan R. Liss.

———. 1999. *Mother Nature: Maternal Instincts and How They Shape the Human Species.* New York: Ballentine Books.

———. 2000. "The Optimal Number of Fathers: Evolution, Demography, and History in the Shaping of Female Mate Preferences." *Annals of the New York Academy of Sciences* 907: 75–96.

———. 2009. *Mothers and Others: The Evolutionary Origins of Human Understanding.* Cambridge, Mass: Harvard University Press.

Hrdy, Sarah B., and George C. Williams. 1983. "Behavioral Biology and the Double Standard." In *Social Behavior of Female Vertebrates,* ed. S.K. Wasser, 3–17. New York: Academic Press.

Isaac, Glynn. 1978. "The Food-Sharing Behavior of Proto-Human Hominids." *Scientific American* 238(4): 90–108.

Isler, Karin, and Carel P. van Schaik. 2012. "How Our Ancestors Broke Through the Gray Ceiling: Comparative Evidence for Cooperative Breeding in Early *Homo.*" *Current Anthropology* 53: S453–65.

Isles, Anthony R., William Davies, and Lawrence S. Wilkinson. 2006. "Genomic Imprinting and the Social Brain." *Philosophical Transactions of the Royal Society B* 361: 2220–37.

James, Wendy. 2008. "Why 'Kinship'? New Questions on an Old Topic." In *Early Human Kinship: From Sex to Social Reproduction*, ed. Nicholas J. Allen et al., 3–20. Oxford: Blackwell.

———. 2017. "'From Lucy to Language: The Archaeology of the Social Brain': An Open Invitation for Social Anthropology to Join the Evolutionary Debate." In *Human Origins: Contributions from Social Anthropology*, ed. Camilla Power et al., 293–318. New York and London: Berghahn.

Jaubert, Jacques, et al. 2016. "Early Neanderthal Constructions Deep in Bruniquel Cave in Southwestern France." *Nature* 534: 111–14.

Jolly, Clifford. 1970. "The Seed Eaters: A New Model of Hominid Differentiation Based on a Baboon Analogy." *Man* 5: 5–27.

———. 2009. "Fifty Years of Looking at Human Evolution." *Current Anthropology* 50: 187–99.

Jones, Clara B. 2005. *Behavioral Flexibility in Primates: Causes and Consequences*. New York: Springer Science + Business Media, Inc.

Jones, Doug. 2000. "Group Nepotism and Human Kinship." *Current Anthropology* 41(5): 779–809.

———. 2011. "The Matrilocal Tribe: An Organization of Demic Expansion." *Human Nature* 22(1–2): 177–200.

Joordens, Josephine C.A., et al. 2009. "Relevance of Aquatic Environments for Hominins: A Case Study from Trinil (Java, Indonesia)." *Journal of Human Evolution* 57: 656–71.

Joordens, Josephine C.A., et al. 2011. "An Astronomically-Tuned Climate Framework for Hominins in the Turkana Basin." *Earth and Planetary Science Letters* 307(1–2): 1–8.

Kaplan, H., K. Hill, J. Lancaster, and A.M. Hurtado. 2000. "A Theory of Human Life History Evolution: Diet, Intelligence and Longevity." *Evolutionary Anthropology* 9: 156–85.

Keverne, Eric B. 1992. "Primate Social Relationships: Their Determinants and Consequences." *Advances in the Study of Behavior* 21: 1–38.

———. 2017. *Beyond Sex Differences: Genes, Brains, and Matrilineal Evolution*. Cambridge: Cambridge University Press.

Keverne, Eric B., et al. 1996a. "Primate Brain Evolution: Genetic and Functional Considerations." *Proceedings of the Royal Society of London B* 263: 689–96.

———. 1996b. "Genomic Imprinting and the Differential Roles of Parental Genomes in Brain Development." *Developmental Brain Research* 92(1): 91–100.

Keverne, Eric B., and James P. Curley. 2008. "Epigenetics, Brain Evolution and Behaviour." *Neuroendocrinology* 29(3): 398–412.

Key, C. A. 1998. *Cooperation, Parental Care and the Evolution of Hominid Social Groups*. Doctoral Dissertation, University College London, University of London.

Key, C. A., and C. Ross. 1999. "Sex Differences in Energy Expenditure in Non-Human Primates." *Proceedings of the Royal Society of London B* 266: 2479–85.

Kingston, John D. 2007. "Shifting Adaptive Landscapes: Progress and Challenges in Reconstructing Early Hominid Environments." *Yearbook of Physical Anthropology* 50: 20–58.

Klein, R.G. 1987. "Problems and Prospects in Understanding How Early People Exploited Animals." In *The Evolution of Human Hunting*, ed. M.H. Nitecki and D.V. Nitecki, 11–45. New York: Plenum Press.

———. 2001. "Southern Africa and Modern Human Origins." *Journal of Anthropological Research* 57(1): 1–16.

Knight, Chris. 2008. "Early Human Kinship Was Matrilineal." In *Early Human Kinship: From Sex to Social Reproduction*, ed. Nicholas J. Allen et al., 61–82. Oxford: Blackwell.

Knight, Chris, and C. Power. 2005. "Grandmothers, Politics, and Getting Back to Science." In *Grandmotherhood: The Evolutionary Significance of the Second Half of Life*, ed. E. Voland, A. Chasiotis, and W. Schiefenhovel, 81–98. New Brunswick and London: Rutgers University Press.

Kostjens, Amanda H. 2008. "The Importance of Kinship in Monkey Society." In *Early Human Kinship: From Sex to Social Reproduction*, ed. Nicholas J. Allen et al., 151–59. Oxford: Blackwell.

Krader, Lawrence. 1979. "Comment on C. Fluehr-Lobban, 'A Marxist Reappraisal of the Matriarchate'." *Current Anthropology* 20(2): 341–59.

Krause, J., et al. 2007. "The Derived FOXP2 Variant of Modern Humans Was Shared with Neanderthals." *Current Biology* 17: 1908–12.

Kroeber, A. L. 1909. "Classificatory Systems of Relationship." *Journal of the Royal Anthropological Institute* 39: 77–84.

———. 1923. *Anthropology*. New York: Harcourt, Brace.

Kuhn, S. L. 1995. *Mousterian Lithic Technology: An Ecological Perspective*. Princeton: Princeton University Press.

Kuhn, Steven L., and Mary C. Stiner. 2006. "What's a Mother to Do? The Division of Labor Among Neanderthals and Modern Humans in Eurasia." *Current Anthropology* 47(6): 953–80.

Kuliukas, Algis V. 2011. "Critique of the Aquatic Ape Hypothesis: Its Final Refutation or Just Another Misunderstanding?" In *Was Man More Aquatic in the Past? Fifty Years After Alister Hardy Waterside Hypotheses of Human Evolution*, ed. Mario Vaneechoutee, Algis Kuliukas, and Marc Verhaegen, 213–25. New York: Bentham Books.

Lalueza-Fox, Carles, et al. 2011. "Genetic Evidence for Patrilocal Mating Behavior Among Neandertal Groups." *Proceedings of the National Academy of Sciences* 108: 250–53.

Lancaster, C. S. 1976. "Women, Horticulture and Society in Sub-Saharan Africa." *American Anthropologist* 78(3): 539–64.

———. 1979. "Battle of the Sexes: A Reply to Karla Poewe." *American Anthropologist* 81(1): 117–19.

Lancaster, Chet S., and Jane Beckman Lancaster. 1978. "On the Male Supremacist Complex: A Reply to Divale and Harris." *American Anthropologist* 80: 115–17.

Lancaster, Jane. 1976. "Sex Roles in Primate Societies." In *Sex Differences*, ed. M.S. Teitelbaum. New York: Doubleday/Anchor Press.

Landau, Misia. 1984. "Human Evolution as Narrative." *American Scientist* 72: 262–68.

Langdon, J. H. 1997. "Umbrella Hypotheses and Parsimony in Human Evolution: A Critique of the Aquatic Ape Hypothesis." *Journal of Human Evolution* 33: 479–94.

Latour, B., and S. C. Strum. 1986. "Human Social Origins: Oh Please, Tell Us Another Story." *Journal of Social and Biological Structures* 9: 169–87.
Laughlin, William S. 1968. "Hunting: An Integrating Biobehavior System and Its Evolutionary Importance." In *Man the Hunter*, ed. Richard B. Lee and Irven DeVore, 304–20. Chicago: Aldine.
Layton, Robert. 2008. "What Can Ethnography Tell Us About Human Social Evolution?" In *Early Human Kinship: From Sex to Social Reproduction*, ed. Nicholas J. Allen et al., 113–27. Oxford: Blackwell.
Layton, Robert, and Sean O'Hara. 2010. "Human Social Evolution: A Comparison of Hunter Gatherer and Chimpanzee Social Organization." In *Social Brain, Distributed Mind*, ed. Robin Dunbar, Clive Gamble, and John Gowlett, 83–113. Oxford: Oxford University Press.
Leacock, Eleanor. 1954. "The Montagnais 'Hunting Territory' and the Fur Trade." *American Anthropological Association Memoir* 78.
———. 1955. "Matrilocality in a Simple Hunting Economy." *Southwestern Journal of Anthropology* 11: 31–47.
———. 1978. "Women's Status in Egalitarian Society: Implications for Social Evolution." *Current Anthropology* 19(2): 225–59.
Ledon-Rettig, Cris C., Christina L. Richards, and Lynn B. Martin. 2012. "Epigenetics for Behavioral Ecologists." *Behavioral Ecology* 24(2): 311–24.
Lee, Richard B., and Irven Devore, eds. 1968. *Man the Hunter*. Chicago: Aldine Publishing Company.
Lee-Thorp, J., J. Thackery, and N. van der Merwe. 2000. "The Hunters and the Hunted Revisited." *Journal of Human Evolution* 39: 565–76.
Lehmann, Julia, Amanda H. Korstjens, and R.I.M. Dunbar. 2007. "Fission-Fusion Social Systems as a Strategy for Coping with Ecological Constraints: A Primate Case." *Evolutionary Ecology* 21: 613–34.
Leonard, William R., and Marcia L. Robertson. 1998. "Comparative Primate Energetics and Hominid Evolution." *American Journal of Physical Anthropology* 102(2): 265–81.
Levi-Strauss, Claude. 1967. *Les Structures Elementaires de la Parente*. 2nd ed. Paris: Mouton.
Linton, Ralph. 1936. *The Study of Man*. New York: Appleton-Century.
Linton, Sally. 1971. "Woman the Gatherer: Male Bias in Anthropology." In *Woman in Perspective: A Guide for Cross-Cultural Studies*, ed. S. E. Jacobs, 9–21. Urbana: University of Illinois Press.
Lippert, Julius. 1931 (orig. 1886–87). *The Evolution of Culture*, trans. G. P. Murdock. New York: Macmillan.
Lorenz, Konrad. 1966. *On Aggression*. London: Metheun.
Lovejoy, Owen C. 1981. "The Origin of Man." *Science* 211: 341–50.
———. 2009. "Reexamining Human Origins in Light of *Ardipithecus ramidus*." *Science* 326: 74e1–8.
Low, Chris. 2017. "Human Physiology, San Shamanic Healing and the 'Cognitive Revolution.'" In *Human Origins: Contributions from Social Anthropology*, ed. Camilla Power et al., 224–47. New York: Berghahn.
Lowenstein, J. M., and A. L. Zihlman. 1980. "The Wading Ape—A Watered Down Version of Human Evolution." *Oceans* 17: 3–6.
Lowie, Robert. 1920. *Primitive Society*. New York: Boni and Liveright.

Lubbock, John. 1873 (orig. 1870). *The Origin of Civilization and the Primitive Condition of Man: Mental and Social Condition of Savages*. New York: D. Appleton.

Mace, R., and M. Pagel. 1994. "Comparative Method in Anthropology." *Current Anthropology* 35: 549–64.

Makarius, Raoul. 1977. "Ancient Society and Morgan's Kinship Theory 100 Years After." *Current Anthropology* 18(4): 709–29.

Malaivijitnond, S., et al. 2007. "Stone Tool Usage by Thai Long-Tailed Macaques (*macaca fascicularis*)." *American Journal of Primatology* 69: 227–33.

Malinowski, Bronislaw 1929. *The Sexual Life of Savages in North-Western Melanesia: An Ethnographic Account of Courtship, Marriage and Family Life Among the Natives of the Trobriand Islands, British New Guinea*. New York: Harcourt Brace.

———. 1956 (orig. 1931). *Marriage: Past and Present: A Debate Between Robert Briffault and Bronislaw Malinowski*, ed. M. F. Ashley Montague. Boston: Porter Sargent.

Manzi, G., F. Mallegni, and A. Ascenzi. 2001. "A Cranium for the Earliest Europeans: Phylogenetic Position of the Hominid from Ceprano, Italy." *Proceedings of the National Academy of Sciences* 98(17): 10011–16.

Marean, Curtis W., et al. 2007. "Early Human Use of Marine Resources and Pigment in South Africa During the Middle Pleistocene." *Nature* 449: 905–8.

Marino, Lori. 2006. "Absolute Brain Size: Did We Throw the Baby Out with the Bathwater?" *Proceedings of the National Academy of Sciences* 103(37): 13563–64.

Marlowe, Frank W. 2004. "Marital Residence Among Foragers." *Current Anthropology* 45: 277–83.

Martin, Emily. 1991. "The Egg and the Sperm: How Science Has Constructed a Romance Based on Stereotypical Male-Female Roles." *Signs* 16(3): 485–501.

Martin, M. Kay. 1969. "South American Foragers: A Case Study in Cultural Devolution." *American Anthropologist* 71: 243–60.

———. 1970. *The Australian Band: A Diachronic Model of Post-Contact Change*. Doctoral Dissertation. Ann Arbor: University Microfilms.

———. 1974. *The Foraging Adaptation: Uniformity or Diversity?* Addison-Wesley Modular Publication No. 56. Reading, Mass: Addison-Wesley.

Martin, M. Kay, and Barbara Voorhies. 1975. *Female of the Species*. New York: Columbia University Press.

Martinez, I., et al. 2004. "Auditory Capacities in Middle Pleistocene Humans from the Sierra de Atapuerca in Spain." *Proceedings of the National Academy of Sciences* 101(27): 9976–81.

Marx, Karl. 1965 (orig. 1857–58). *Pre-Capitalist Economic Formations: Karl Marx*, trans. J. Cohen, New York: International Publishers.

Maslin, Mark A., and Martin H. Trauth. 2009. "Plio-Pleistocene East African Pulsed Climate Variability and Its Influence on Early Human Evolution." In *The First Humans—Origins of the Genus Homo*, ed. F. E. Grine, R. E. Leakey, and J. G. Fleagle, 151–58. Springer Science.

Maslin, Mark A., Susanne Schultz, and Martin H. Trauth. 2015. "A Synthesis of the Theories and Concepts of Early Human Evolution." *Philosophical Transactions of the Royal Society B* 370: 20140064.

Maslin, M., et al. 2015. "East African Climate Pulses and Early Human Evolution." *Quaternary Science Reviews* 101: 1–17.

Mavalwala, Jamshed. 1976. "Comment on John Hartung, et al., 'On Natural Selection and the Inheritance of Wealth'." *Current Anthropology* 17(4): 617–18.
Mayr, Ernst. 1950. "Taxonomic Categories of Fossil Hominids." *Cold Spring Harbor Symposia on Quantitative Biology* 15: 109–18.
McBrearty, Sally, and Alison S. Brooks. 2000. "The Revolution That Wasn't: A New Interpretation of the Origin of Modern Human Behavior." *Journal of Human Evolution* 39: 453–63.
McLennan, John. 1865. *Primitive Marriage*. Edinburgh: Adam and Charles Black.
McHenry, Henry M. 1994. "Behavioral Ecological Implications of Early Hominid Body Size." *Journal of Human Evolution* 27: 77–87.
McHenry, Henry M., and Katherine Coffing. 2000. "Australopithecus to *Homo*: Transformations in Body and Mind." *Annual Review of Anthropology* 29: 125–46.
Mery, Frederic, and James G. Burns. 2010. "Behavioral Plasticity: An Interaction Between Evolution and Experience." *Evolutionary Ecology* 24(3): 571–83.
Mills, D., and M. T. Huber. 2005. "Anthropology and the Educational 'Trading Zone': Disciplinarity, Pedagogy, and Professionalism." *Arts and Humanities in Higher Education* 4(1): 9–32.
Milton, Katherine. 1999. "A Hypothesis to Explain the Role of Meat-Eating in Human Evolution." *Evolutionary Anthropology* 8(1): 11–21.
Mithen, S. 1996. *The Prehistory of the Mind: A Search for the Origins of Art, Science, and Religion*. London: Thames and Hudson.
Moczek, Armin P., et al. 2011. "The Role of Developmental Plasticity in Evolutionary Innovation." *Proceedings of the Royal Society of Britain B* 278: 2705–13.
Morgan, Elaine. 1972. *The Descent of Woman*. London: Souvenir Press.
———. 1982. *The Aquatic Ape*. London: Souvenir Press.
———. 1990. *The Scars of Evolution*. Oxford: Oxford University Press.
———. 1997. *The Aquatic Ape Hypothesis*. London: Souvenir Press.
Morgan, Lewis Henry. 1870. *Systems of Consanguinity and Affinity of the Human Family*. Washington, DC: Smithsonian Institution.
———. 1907 (orig. 1877). *Ancient Society*. London: Macmillan.
Morgan, T. J. H., et al. 2015. "Experimental Evidence for the Co-Evolution of Hominin Tool-Making Teaching and Language." *Nature Communications* 6, No. 6029.
Moss, M. L. 1993. "Shellfish, Gender, and Status on the Northwest Coast: Reconciling Archaeological, Ethnographic, and Ethnohistorical Records of the Tlingit." *American Anthropologist* 95: 631–52.
Murdock, George P. 1937. "Comparative Data on the Division of Labor by Sex." *Social Forces* XV: 551–53.
———. 1949. *Social Structure*. New York: The Free Press.
———. 1959. *Africa: Its Peoples and Their Culture History*. New York: McGraw-Hill.
———. 1967. *Ethnographic Atlas*. Pittsburgh: University of Pittsburgh Press.
Nadeau, Joseph H. 2017. "Do Gametes Woo? Evidence for Their Nonrandom Union at Fertilization." *Genetics* 207(2): 369–87.
Narroll, Raoul. 1965. "Galton's Problem: The Logic of Cross-Cultural Research." *Social Research* 32: 428–51.
Nicholas, George P. 1998. "Wetlands and Hunter-Gatherers: A Global Perspective." *Current Anthropology* 19(5): 720–31.

Nimchinsky, E. A., et al. 1999. "A Neuronalmorphologic Type Unique to Humans and Great Apes." *Proceedings of the National Academy of Science. U.S.A.* 96(9): 5268–73.

Nowak, Martin, Corina E. Tarnita, and Edward O. Wilson. 2010. "The Evolution of Eusociality." *Nature* 466: 1057–62.

Nowak, Martin, and Roger Highfield. 2011. *Super Cooperators: Altruism, Evolution, and Why We Need Each Other to Succeed.* New York: Free Press.

Oberg, Kalervo. 1955. "Types of Social Structure Among Lowland Tribes of South and Central America." *American Anthropologist* 57: 472–88.

O'Connell, J.F., K. Hawkes, and N.G. Blurton Jones. 1999. "Grandmothering and the Evolution of *Homo erectus*." *Journal of Human Evolution* 36: 461–85.

O'Connell, J.F., K. Hawkes, K.D. Lupo, and N.G. Blurton Jones. 2002. "Male Strategies and Plio-Pleistocene Archaeology." *Journal of Human Evolution* 43: 831–72.

Opie, Kit, and Camilla Power. 2008. "Grandmothering and Female Coalitions: A Basis For Matrilineal Priority?" In *Early Human Kinship: From Sex to Social Reproduction,* ed. Nicholas J. Allen et al., 168–86. Oxford: Blackwell.

Orians, Gordon H. 1969. "On the Evolution of Mating Systems in Birds and Mammals." *The American Naturalist* 103(934): 589–603.

Otterbein, Keith. 1968. "Internal War. A Cross-Cultural Study." *American Anthropologist* 70: 277–89.

Otterbein, K. F., and C. S. Otterbein. 1965. "An Eye for an Eye, a Tooth for a Tooth: A Cross-Cultural Study of Feuding." *American Anthropologist* 67: 1470–82.

Pappu, S. 2011. "Early Pleistocene Presence of Acheulean Hominins in South India." *Science* 331: 1596–99.

Parker, Seymour, and Hilda Parker. 1979. "The Myth of Male Superiority: Rise and Demise." *American Anthropologist* 81(2): 289–309.

Passingham, Richard E. 2002. "The Frontal Cortex: Does Size Matter?" *Nature Neuroscience* 5: 190–92.

Pavard, Samuel, et al. 2007a. "The Influence of Maternal Care in Shaping Human Survival and Fertility." *Evolution* 61–12: 2801–10.

———. 2007b. "The Effect of Maternal Care on Child Survival: A Demographic, Genetic and Evolutionary Perspective." *Evolution* 61–65: 1153–61.

Pawlowskil, Boguslaw, C.B. Lowen, and R.I.M. Dunbar. 1998. "Neocortex Size, Social Skills, and Mating Success in Primates." *Behaviour* 135(3): 357–68.

Pennisi, Elizabeth. 2016. "Do Genomic Conflicts Drive Evolution?" *Science* 353(6297): 334–35.

Pinker, Steven. 1994. *The Language Instinct.* New York: HarperCollins Publishers.

Plavcan, J. Michael. 2012. "Body Size, Size Variation, and Sexual Size Dimorphism in Early *Homo*." *Current Anthropology* 53(S6): S409–23.

Poewe, Karla O. 1979. "Women, Horticulture, and Society in Sub-Saharan Africa: Some Comments." *American Anthropologist* 81(1): 115–17.

Pontzer, Herman. 2012. "Ecological Energetics in Early *Homo*." *Current Anthropology* 53(S6): S346–S358.

Potts, Richard. 1996. "Evolution and Climate Variability." *Science* 273: 922–23.

———. 1998. "Environmental Hypotheses of Hominin Evolution." *American Journal of Physical Anthropology* 27: 93–136.

———. 2013. "Hominin Evolution in Settings of Strong Environmental Variability." *Quaternary Science Reviews* 73: 1–13.
Powell, Adam, Steven Shennan, and Mark G. Thomas. 2009. "Late Pleistocene Demography and the Appearance of Modern Behavior." *Science* 324: 1298–1301.
Power, Camilla, Morna Finnegan, and Hilary Callan, eds. 2017. *Human Origins: Contributions from Social Anthropology*. New York and Oxford: Berghahn.
Rabinovich, Rivka, Sabina Gaudzinski-Windhauser, and Naama Goren-Inbar. 2008. "Systematic Butchering of Fallow Deer (*Dama*) at the Early Middle Pleistocene Acheulian Site of Geshner Benot Ya'aqov (Israel)." *Journal of Human Evolution* 54(1): 134–49.
Radcliffe-Brown, A. R., and Daryll Forde. 1950. "Introduction." In *African Systems of Kinship and Marriage*, ed. A. R. Radcliffe-Brown and Daryll Forde, 1–85. London: Oxford University Press for the International African Institute.
Ralls, Katherine. 1976. "Mammals in Which Females are Larger than Males." *Quarterly Review of Biology* 51: 245–76.
Reader, Simon M., and Kevin N. Laland. 2002. "Social Intelligence, Innovation, and Enhanced Brain Size in Primates." *Proceedings of the National Academy of Science* 99(7): 4436–41.
Reed, Kaye E. 1997. "Early Hominid Evolution and Ecological Change Through the African Plio-Pleistocene." *Journal of Human Evolution* 32: 289–322.
Renfrew, Colin. 2007. *Prehistory: The Making of the Human Mind*. London: Weidenfeld and Nicholson.
Richards, Audrey I. 1950. "Some Types of Family Structure Amongst the Central Bantu." In *African Systems of Kinship and Marriage*, ed. A. R. Radcliffe-Brown and Daryll Forde, 207–51. London: Oxford University Press for the International African Institute.
Richards, Christina L., Oliver Bossdorf, and Massimo Pigliucci. 2010. "What Role Does Heritable Epigenetic Variation Play in Phenotypic Evolution?" *BioScience* 60(3): 232–37.
Rilling, James K., et al. 2012. "Differences Between Chimpanzees and Bonobos in Neural Systems Supporting Social Cognition." *Social Cognitive and Affective Neuroscience* 7: 369–79.
Robinson, Gene E., and Andrew B. Barron. 2017. "Epigenetics and the Evolution of Instincts." *Science* 356(6333): 26–27.
Rodseth, Lars, Richard W. Wrangham, Alisa M. Harrigan, and Barbara B. Smuts. 1991. "The Human Community as a Primate Society." *Current Anthropology* 32(3): 221–54.
Ross, Alexander. 1904. "Adventures of the First Settlers on the Oregon or Columbia River; Being a Narrative of an Expedition Fitted Out by John Jacob Astor, to Establish the 'Pacific Fur Company'; With an Account of the Indian Tribes on the Coast of the Pacific." In *Early Western Travels*, ed. Ruben Thwaites, Vol. 7. Cleveland: Arthur H. Clark Co.
Ryan, Christopher, and Cacilda Jetha. 2010. *Sex at Dawn*. New York: Harper Collins.
Sacks, Karen. 1975. "Engels Revisited: Women, the Organization of Production, and Private Property." In *Toward an Anthropology of Women*, ed. Rayna R. Reiter. New York: Monthly Review.

Sahlins, Marshall. 1960. "The Origin of Society." *Scientific American* 203(1): 76–87.
———. 1961. "The Segmentary Lineage: An Organization of Predatory Expansion." *American Anthropologist* 63: 322–43.
Sanday, Peggy R. 1973. "Toward a Theory of the Status of Women." *American Anthropologist* 75(5): 1682–1700.
———. 1974. "Female Status in the Public Domain." In *Woman, Culture and Society*, ed. Michelle Zimbalist and Louise Lamphere. Stanford: Stanford University Press.
Sawaguchi, T. 1992. "The Size of the Neocortex in Relation to Ecology and Social Structure in Monkeys and Apes." *Folia Primatologica* 58: 131–45.
Schagatay, Erika. 2011. "Human Breath-Hold Diving Ability Suggests a Selective Pressure for Diving During Human Evolution." In *Was Man More Aquatic in the Past? Fifty Years After Alister Hardy Waterside Hypotheses of Human Evolution*, ed. Mario Vaneechoutee, Algis Kuliukas, and Marc Verhaegen, 120–47. New York: Bentham Books.
Schepartz, L. A. 1993. "Language and Modern Human Origins." *Yearbook of Physical Anthropology* 36: 91–126.
Schillaci, Michael A. 2008. "Primate Mating Systems and the Evolution of Neocortex Size." *Journal of Mammalogy* 89(1): 58–63.
Schlegel, Alice. 1972. *Male Dominance and Female Autonomy: Domestic Authority in Matrilineal Societies*. HRAF Press.
Schneider, David M. 1961. "Introduction: The Distinctive Features of Matrilineal Descent Groups." In *Matrilineal Kinship*, ed. David M. Schneider and Kathleen Gough, 1–35. Berkeley and Los Angeles: University of California Press.
Schneider, David M., and Kathleen Gough, eds. 1961. *Matrilineal Kinship*. Berkeley and Los Angeles: University of California Press.
Semendeferi, K., et al. 2002. "Humans and Great Apes Share a Large Frontal Cortex." *Nature Neuroscience* 5(3): 272–76.
Semenov, Yu. I. 1965. "The Doctrine of Morgan, Marxism and Contemporary Ethnography." *Soviet Anthropology and Archeology* 4: 3–15.
Service, Elman. 1962. *Primitive Social Organization*. New York: Random House.
Shea, John J. 2006. "Comment" on S. L. Kuhn and S. C. Stiner, "What's a Mother to Do? The Division of Labor Among Neanderthals and Modern Humans in Eurasia." *Current Anthropology* 47(6): 968.
Shennan, Stephan. 2011. "Property and Wealth Inequality as Cultural Niche Construction." *Philosophical Transactions of the Royal Society B* 366: 918–26.
Sherman, Paul, Eileen Lacey, Hudson Reeve, and Laurent Keller. 1995. "Forum: The Eusociality Continuum." *Behavioral Ecology* 6: 102–8.
Slocum, S. 1975. "Woman the Gatherer: Male Bias in Anthropology." In *Toward an Anthropology of Women*. New York: Monthly Review Press.
Smaers, J. B., and C. Soligo. 2013. "Brain Reorganization, Not Relative Brain Size, Primarily Characterizes Anthropoid Brain Evolution." *Proceedings of the Royal Society B* 280: 20130269. doi.org/10.1098/rspb.2013.0269.
Small, Meredith. 1993. *Female Choices: Sexual Behavior of Female Primates*. Ithaca, New York: Cornell University Press.
———, ed. 1984. *Female Primates: Studies by Women Anthropologists*. New York: Alan R. Liss.

Smith, Alex R., et al. 2015. "The Significance of Cooking for Early Hominin Scavenging." *Journal of Human Evolution* 84: 62–70.
Smith, Eric Alden, et al. 2010. "Wealth Transmission and Inequality Among Hunter-Gatherers." *Current Anthropology* 51(1): 19–34.
Smith, Kerry, Susan C. Alberts, and Jeanne Altmann. 2003. "Wild Females Bias Their Behaviour Toward Paternal Half-Sisters." *Proceedings of the Royal Society B* 270: 503–10.
Smuts, B. B. 1987. "Gender, Aggression and Influence." In *Primate Societies*, ed. B.B. Smuts et al., 400–12. Chicago: University of Chicago Press.
Snell-Rood, Emilie C. 2013. "An Overview of the Evolutionary Causes and Consequences of Behavioural Plasticity." *Animal Behavior* 85(5): 1004–11.
Snell-Rood, E. C., et al. 2010. "Toward a Population Genetic Framework of Developmental Evolution: The Costs, Limits, and Consequences of Phenotypic Plasticity." *Bioessays* 32: 71–81.
Snow, D. R. 1995. "Migration in Prehistory: The Northern Iroquois Case." *American Antiquity* 60: 59–79.
———. 1996. "More on Migration in Prehistory: Accommodating New Evidence in the Northern Iroquois Case." *American Antiquity* 61: 791–96.
Soffer, Olga. 2006. "Comment" on S. L. Kuhn and S. C. Stiner, "What's a Mother to Do? The Division of Labor Among Neanderthals and Modern Humans in Eurasia." *Current Anthropology* 47(6): 968–69.
Spencer, Herbert. 1896 (orig.1876). *Principles of Sociology*. New York: D. Appleton.
Stanley, S. M. 1992. "An Ecological Theory for the Origin of *Homo*." *Paleobiology* 18: 237–57.
Steele, Teresa E. 2010. "A Unique Hominin Menu Dated to 1.95 Million Years Ago." *Proceedings of the National Academy of Sciences* 107(24): 10771–72.
Steudel-Numbers, Karen. 2006. "Energetics in *Homo erectus* and Other Early Hominins: The Consequences of Increased Lower-Limb Length." *Journal of Human Evolution* 51: 445–53.
Steward, Julian. 1946. *Handbook of South American Indians*. Bureau of American Ethnology Bulletin No. 143, Vol. 1. Washington, DC.
———. 1955. *Theory of Culture Change*. Urbana: University of Illinois Press.
Steward, Julian, and Louis Faron. 1959. *Native Peoples of South America*. New York: McGraw-Hill.
Stewart, Kathlyn M. 1994. "Early Hominid Utilisation of Fish Resources and Implications for Seasonality and Behaviour." *Journal of Human Evolution* 27: 229–45.
Stiner, Mary C. 2002. "Carnivory, Coevolution, and the Geographic Spread of the Genus *Homo*." *Journal of Archaeological Research* 10(1): 1–63.
Stringer, Chris, and Julia Galway-Witham. 2017. "Palaeoanthropology: On the Origin of Our Species." *Nature* 546: 212–14.
Symons, Donald. 1979. *The Evolution of Human Sexuality*. Oxford: Oxford University Press.
Tanner, Nancy. 1981. *On Becoming Human*. New York: Cambridge University Press.
Tiger, Lionel. 1969. *Men in Groups*. New York: Vintage Books.
———. 1970. "The Possible Biological Origins of Sexual Discrimination." *Impact of Science on Society* 20: 29–44.

Tiger, Lionel, and Robin Fox. 1971. *The Imperial Animal*. New York: Holt, Rinehart and Winston.

Tobias, Phillip V. 1998. "Water and Human Evolution." *Out There* 3: 38–44.

———. 2011. "Revisiting Water and Hominin Evolution." In *Was Man More Aquatic in the Past? Fifty Years After Alister Hardy Waterside Hypotheses of Human Evolution*, ed. Mario Vaneechoutee, Algis Kuliukas, and Marc Verhaegen, 3–15. New York: Bentham Books.

Toth, N., and K. D. Schick. 1993. "Early Stone Industries and Inferences Regarding Language and Cognition." In *Tools, Language and Cognition in Human Evolution*, ed. K. Gibson and T. Ingold, 346–62. Cambridge, UK: Cambridge University Press.

Trigger, B. G. 1976. *The Children of Ataensic: A History of the Huron People to 1660*. Montreal: McGill-Queen's University Press.

———. 1978. "Iroquoian Matriliny." *Pennsylvania Archeologist* 48(1–2): 55–65.

Trivers, R. L. 1971. "The Evolution of Reciprocal Altruism." *Quarterly Review of Biology* 46: 35–57.

———. 1972. "Parental Investment and Sexual Selection." In *Sexual Selection and the Descent of Man*, ed. B. Campbell, 136–79. Chicago: Aldine Publishing Company.

———. 1974. "Parent-Offspring Conflict." *American Zoologist* 14: 249–64.

Tylor, Edward B. 1889. "On a Method of Investigating the Development of Institutions Applied to the Laws of Marriage and Descent." *Journal of the Royal Anthropological Institute of Great Britain and Ireland* 18: 245–69.

Ungar, Peter S. 2012. "Dental Evidence for the Reconstruction of Diet in African Early *Homo*." *Current Anthropology* 53(S6): S318–29.

———. 2017. *Evolution's Bite: A Story of Teeth, Diet and Human Origins*. Princeton, NJ: Princeton University Press.

van den Berghe, Pierre. 1979. *Human Family Systems*. Prospect Heights, Illinois: Waveland Press.

van den Berghe, Pierre L., and David P. Barash. 1977. "Inclusive Fitness and Human Family Structure." *American Anthropologist* 79: 809–23.

van der Blick, Alexander M. 2016. "Demystifying the Demise of Paternal Mitochondrial DNA." *Science* 353(6297): 351–52.

Vaneechoute, Mario, Algis Kuliukas, and Marc Verhaegen, eds. 2011. *Was Man More Aquatic in the Past? Fifty Years After Alister Hardy Waterside Hypotheses of Human Evolution*. New York: Bentham Books.

Van Velzen, H.U.E. Thoden, and W. van Wetering. 1960. "Residence, Power Groups and Intra-Societal Aggression." *International Archives of Ethnography* 49: 169–200.

Vekua, Abesalom, et al. 2002. "A New Skull of Early *Homo* from Dmanisi, Georgia." *Science* 7(5578): 85–89.

von Bertalanffy, Ludwig. 1968. *General Systems Theory*. New York: George Braziller.

Vrba, Elizabeth. 1995. "The Fossil Record of African Antelopes (Mammalia, Bovidea) in Relation to Human Evolution and Paleoclimate." In *Paleoclimate and Evolution with Emphasis on Human Origins*, ed. E. Vrba, G. Denton, L. Burckle, and T. Patridge, 35–424. New Haven, CT: Yale University Press.

Washburn, Sherwood L., ed. 1961. *The Social Life of Early Man*. Chicago: Aldine Publishing Company.
Washburn, Sherwood, and C. S. Lancaster. 1968. "The Evolution of Hunting." In *Man the Hunter*, ed. Richard D. Lee and Irven Devore, 293–303. Chicago: Aldine Publishing Company.
Watson-Franke, Maria-Barbara. 1992. "Masculinity and the 'Matrilineal Puzzle'." *Anthropos* 87: 45–88.
———. 1995. "Revisionism or the Recovery of Matrilineal Women's Centrality? A Reply to Bolyanatz." *Anthropos* 90: 582–85.
Weaver, Anne H., and Erik Trinkaus. 2005. "Reciprocal Evolution of the Cerebellum and Neocortex in Fossil Humans." *Proceedings of the National Academy of Science, U.S.A.* 12(10): 3576–80.
Wengrow, David, and David Graeber. 2015. "Farewell to the 'Childhood of Man': Ritual, Seasonality, and the Origins of Inequality." *Journal of the Royal Anthropological Institute* 21: 597–619.
Westermarck, E. 1894 (orig.1891). *The History of Human Marriage*. New York: Macmillan.
White, Leslie A. 1949. *The Science of Culture*. New York: Grove Press.
———. 1959. *The Evolution of Culture*. New York: McGraw-Hill.
White, T.D., G. Suwa, and B. Asfaw. 1994. "Australopithecus ramidus, a New Species of Early Hominid from Aramis, Ethiopia." *Nature* 371: 306–12.
Whittaker, John C., and Grant McCall. 2001. "Handaxe-Hurling Hominids: An Unlikely Story." *Current Anthropology* 42: 566–72.
Widdig, A., et al. 2001. "Paternal Relatedness and Age Proximity Regulate Social Relationships Among Adult Female Rhesus Macaques." *Proceedings of the National Academy of Science USA* 98: 13769–73.
Wilkins, Jayne, Benjamin J. Schoville, Kyle S. Brown, and Michael Chazan. 2012. "Evidence for Early Hafted Hunting Technology." *Science* 338(6109): 942–46.
Wilkins, Jon F., and David Haig. 2003. "What Good Is Genomic Imprinting: The Function of Parent-Specific Gene Expression." *Nature Reviews/Genetics* 4: 1–10.
Williams, Tess. 2011. "Just Add Water: The Aquatic Ape Story in Science." In *Was Man More Aquatic in the Past? Fifty Years After Alister Hardy Waterside Hypotheses of Human Evolution*, ed. Mario Vaneechoutee, Algis Kuliukas, and Marc Verhaegen, 199–212. New York: Bentham Books.
Wilson, Edward O. 1975. *Sociobiology: The New Synthesis*. Cambridge, Mass: Harvard University Press.
———. 1978. *On Human Nature*. Cambridge, Mass: Harvard University Press.
———. 2012. *The Social Conquest of Earth*. New York: Liveright Publishing Company.
———. 2014. *The Meaning of Human Existence*. New York: Liveright Publishing Company.
Wittenberger, James F. 1980. "Group Size and Polygamy in Social Mammals." *The American Naturalist* 113(2): 197–222.
Wolpoff, M. H. 1989. "Multiregional Evolution: The Fossil Alternative to Eden." In *The Human Revolution: Behavioural and Biological Perspectives in the Origins*

of Modern Humans, ed. P. Mellars and C. Stringer, 62–108. Edinburgh: Edinburgh University Press.

Wrangham, Richard. 1979. "On the Evolution of Ape Social Systems." *Biology and Social Life: Social Sciences Information* 18(3): 336–68.

———. 1980. "An Ecological Model of Female-Bonded Primate Groups." *Behaviour* 75: 262–300.

———. 2009. *Catching Fire: How Cooking Made Us Human.* New York: Basic Books.

Wrangham, Richard, and Dale Peterson. 1996. *Demonic Males: Apes and the Origins of Human Violence.* New York: Houghton Mifflin Company.

Wright, T.F., et al. 2010. "Behavioral Flexibility and Species Invasions: The Adaptive Flexibility Hypothesis." *Ethology, Ecology and Evolution* 22: 393–404.

Wynn, Thomas. 1993. "Two Developments in the Mind of Early *Homo*." *Journal of Anthropological Archaeology* 12(3): 299–322.

Yellen, J. E., et al. 1995. "A Middle Stone Age Worked Bone Industry from Katanda, Upper Semliki Valley, Zaire." *Science* 268: 553–56.

Zhou, W., et al. 2005. "Discrete Hierarchical Organization of Social Group Sizes." *Proceedings of the Royal Society B* 272: 439–44.

Index

Aberle, David, 194, 202, 207, 209
Aiello, Leslie, 87, 92, 145
Alexander, Richard, 5, 34
Allen, Nicholas, 190–91
Allman, J. M., 141
alloparenting, 31, 56–57, 58, 60–62
  encephalization and, 93, 181
  energetics and, 63–65, 86, 88, 91, 123, 147–48, 152, 156, 174, 181–82
  inclusive fitness and, 174
  subsistence and, 88
anatomically modern humans (AMH)
  African dispersal of, 128
  antiquity of, 10, 129
  aquatic revolution and, 103–5
  labor division and, 111
  sapient paradox, 128–29
  symbolic revolution and, 128, 145, 160
ancestral female kin group hypothesis, 84–96, 232–33
  alloparenting and, 56–57, 92–93, 230–32
  energetics and, 92–93, 181–82
  imprinted genes and, 185–87, 231
  provisioning and, 86–90, 92, 230
ancestral male kin group hypothesis, 35–41, 228–29 (*see also* Pan-genesis theory)
  chimpanzee avatars for, 40, 46–47, 66, 231
  fossil record and, 41, 47
  multilevel social systems and, 40, 48, 176–77
  pair bonding and, 39, 176, 178–79
  sexual contract and, 39
  shortcomings of, 41–48, 177–82
anisogamy, 23–30
  complementarity and, 29–30
  conception and, 24–25, 236n1

  inclusive fitness and, 31, 58–62, 63–64
  parasitism and, 27–29
  reproductive variance and, 171–72
  sexual stereotypes and, 25–26
Anton, Susan, 87, 99
aquatic adaptations
  antiquity of, 77, 80–83, 88–90
  dietary lipids and, 82, 90
  evolutionary significance of, 97–103
aquatic ape hypothesis, 79–80
aquatic revolution theory, 103–5
  critique of, 105–10
Ardipithecus ramidus, 10, 32, 47, 66, 76, 94, 143
Ardrey, Robert, 62, 70
Ashley, Gail, 77
australopithecines, 10, 32, 41, 47–48, 72, 75–76, 82, 91, 93–94, 136
  A. afarensis, 85–86, 93
  A. africanus, 70
  Paranthropus boisei, 73, 76, 82, 85, 92

Bachofen, Johann, 158, 162
Barash, David, 171–72, 174
Barnard, Alan, 138, 146, 160–61
Barron, Andrew, 241n7
behavioral plasticity, 8, 18, 20, 41–42, 44, 68, 189–90, 230, 232–33
  epigenetic traits and, 124–27
  instincts and, 241n7
  primate behaviors and, 126–27
Bickerton, Derek, 160
bilateral kinship
  conditions favoring, 218–22
  primeval origins of, 173, 219–21
  unilineal origins of, 221–22
Binford, Lewis, 103
Boas, Franz, 162
Boaz, Noel, 12, 13, 99, 102, 133–34, 144, 242n13

270 • *Index*

Borgerhoff-Mulder, Monique, 174, 243n8
Boserup, Ester, 216
Bribiescas, Richard, 41
Briffault, Robert, 162
Broadhurst, C. Leigh, 82–83, 89
Brodman, K., 137, 140
Brooks, Alison, 106, 129
Broom, Robert, 70
Brown, Judith, 168

Cachel, Susan, 103–4
Callan, Hilary, 3
Calvin, William, 160
Carillo-Reed, Luis, 141
Chapais, Bernard, 35–36, 39–40, 44–45, 159, 175–80, 187
Chapman, Colin, 148
Chapman, Lauren, 148
Ciochon, Russell, 12, 99, 102, 133–34
Clancy, Kelly, 18
Clark, J. Desmond, 78
climate change effects on
    aquatic resources, 103, 109–10
    hominin migrations, 78, 96
    hominin speciation, 78
    mammalian herbivores, 109
    Plio-Pleistocene ecology, 32–33, 71, 75–78
cooking hypothesis, 83–84, 116–17
cooperative breeding, 31, 56–57, 64, 174, 182. *See also* alloparenting
Copeland, Sandi, 41, 47–48
Copes, Lynn, 134
cuckoldry
    female sexuality and, 39
    matriliny and, 28, 172
Curley, James, 185

Daly, M., 29
Dart, Raymond, 70
Darwin, Charles, 4, 71
Davies, William, 184
Dawkins, Richard, 6, 27–28, 55, 235n4
de Ruiter, Jan, 146–47
Devore, Irven, 36
deWaal, Frans, 66, 128, 137, 142
Diamond, Jared, 131

Dickeman, Mildred, 173, 193
Dickens, Thomas, 125
dietary breadth, 103–10, 112, 116–17, 230
dietary lipids, 80–83, 90, 96, 116, 230
Divale, William, 165–67, 194–99
division of labor
    cooking role in, 83–84
    cooperative breeding and, 64
    dietary breadth and, 111–12
    evolutionary trends in, 224–25
    kinship and, 194, 211–12
    matricentric family and, 44, 64
    nuclear family and, 34, 151
    sexual dimorphism and, 164–65
Dobzhansky, T., 5
Dominguez-Rodrigo, M., 77
Dublin, Holly, 64
Dunbar, Robin, 6, 93, 137–38, 145–47, 160–61
Dunbar's number, 146

ecological constraints model, 148–49
egalitarian societies, 167
Ellison, Peter, 41
Ember, Carol, 194
Ember, Melvin, 194
Emlen, Stephen, 49
encephalization and intelligence, 127–43
energetics, 87–88, 92–93, 151–56
Engels, Frederick, 158, 161–62
epigenesis, 7–8, 125–26
Erlandson, Jon, 101–2, 113
eusociality, 7, 56–57, 72, 127
extended evolutionary synthesis, 7

female bonding, 52–54, 62, 64, 147, 231–32
female choice, 48–53, 55, 61, 174–5
female feeding strategies
    ecological variables and, 48–53
    inclusive fitness and, 58–62, 90
    social organization and, 48–53
female philopatry
    female feeding strategies and, 44, 50–51, 181
    human trait complexes and, 38–39

imprinted genes and, 185–87
primate patterns of, 44, 50–51
primeval origins of, 54–55, 58, 185–87, 232–33
Finlayson, Clive, 11, 13, 76, 101, 105–6, 109, 112, 117, 130–31, 134, 144, 149
Finnegan, Morna, 167
fire use, 99, 116–17
fission-fusion, 120, 148–49, 190, 198, 200, 232
Foley, R. A., 36, 87
Fortes, Meyer, 162
Fox, Robin, 47, 55, 62, 175
fraternal interest groups
    internal warfare and, 46, 194–95, 212, 215
    male inclusive fitness and, 38, 41
    multilevel societies and, 40, 46

Galton's problem, 65–68, 170
Gamble, Clive, 138, 145–46
Gaudzinski-Windhauser, Sabina, 135
general systems theory, 153–55, 191
Ghiglieri, Michael, 36, 38, 44
Gibbons, Ann, 77
Gintis, Herbert, 190
Goodenough, Ward, 192
Goren-Inbar, Naama, 135
Gough, Kathleen, 194, 202, 207–8
Gowlett, John, 181
Graeber, David, 129
grandmother hypothesis, 57. *See also* alloparenting
gray ceiling theory, 93, 152, 178, 232
Gray, Peter, 41
Greuter, Cyril, 36–37, 40, 48
Gusinde, Martin, 221

Haig, David, 184–85
Hamilton, Marcus, 146, 191, 193, 211
Hamilton, W. D., 5
Hamilton's rule, 5, *See also* kin selection theory
Harari, Yuval, 129–30
Hardy, Alister, 79
Harris, Marvin, 161–62, 165–67
Hart, John, 197–98, 201

Hartung, John, 171, 174
Hebb's rule, 141
Henry, Amanda, 106
Herculano-Houzel, Suzana, 140–41, 152
Hoberg, Eric, 99
Holden, Claire, 210
Holliday, Trenton, 103
Holloway, Ralph, 136
Homo antecessor, 108
Homo erectus, 10–11, 69–96
    alloparenting and, 88, 91–93, 181
    aquatic resources and, 80–83, 88–90, 95–96, 102–3, 105
    cranial pachyostosis and, 133–34
    dietary breadth and, 116
    encephalization and, 91–93
    hunting and, 72–75, 95–96
    language and, 134–35, 161
    migrations and, 96–97, 149–50
    scavenging and, 74, 100, 104, 133
    single species theory and, 11–12
    tool-making and, 73–74, 100, 134–35
Homo ergaster, 75, 82, 108
Homo habilis, 10, 69, 72, 82–83, 102, 135–36, 144, 160
Homo heidelbergensis, 10, 73, 100, 108, 132, 136, 160
Homo helmei, 129
Homo naledi, 73, 136
Homo neanderthalensis, 10
    dietary breadth and, 103–8
    disappearance of, 130–31
    labor division and, 111–12
    modern behavior and, 129–32, 160–61
Homo rhodesiensis, 10, 73
Homo sapiens sapiens, 10, 15. *See also* anatomically modern humans
Hrdy, Sarah B., 29, 43, 56–57, 60–61, 63, 123–24
Huber, M. T., 3
human biogram, 2, 7–8, 20, 170–75
    Galton's problem and, 66–67
    plasticity model of, 20, 68, 188
    prescriptive model of, 2, 48, 66, 170–73, 211, 229

hunting hypothesis, 2–3, 16, 70–72, 98, 143–44, 230
  critiques of, 72–75, 83–84, 98–100, 113

imprinted genes, 182–87
  inclusive fitness and, 185–86
  mosaic brain evolution and, 183–85
  philopatry and, 185, 187
inclusive fitness, 5, 7, 17, 44
  female strategies for, 31, 38, 43, 53, 58, 62, 94, 171, 174–75, 213–14
  imprinted genes and, 185–86
  kinship organization and, 213–14
  male strategies for, 14, 31, 38, 53, 60–62, 95, 171, 175, 205–6, 213–14
infanticide, 123
innate male dominance, 164–70
  male supremacist complex and, 165–67
intergenerational wealth transfer, 174, 202, 206, 212, 243n8
Isaac, Glynn, 72, 161
Isler, Karin, 93, 152
Isles, Anthony, 185

James, Wendy, 181
Jetha, Cacilda, 30, 121–22, 124, 240n4
Jolly, Clifford, 35, 45, 159
Jones, Clara, 126
Jones, Doug, 189–90, 192, 198–99
Joordens, Josephine, 78

Kaas, Jon, 140–41
Kaplan, H., 88
Keverne, Eric, 183–87
Key, Cathy, 87, 91
killer-ape hypothesis, 70
Kimbel, William, 134
Kingston, John, 77
kin selection theory, 5, 39, 189
Knight, Chris, 163–64, 181
Kroeber, A. L., 162
Kuhn, Steven, 103–5, 111–12, 151
Kuliukas, Algis, 80

Laland, Kevin, 138, 147
Lalueza-Fox, Carles, 237n6
Lancaster, Chet, 166–67, 169–70, 209
Lancaster, Jane, 166–67, 169
Langdon, J. H., 80
language
  early hominins and, 132–33, 143
  instincts and, 139
  neocortex size and, 138–39
  tool-making and, 134–36
Layton, Robert, 110
Leacock, Eleanor, 168, 170, 209
Lee, Richard B., 36
Lee, P. C., 36, 87
Levi-Strauss, Claude, 175
Linton, Sally, 44
Lorenz, Konrad, 70
Lovejoy, Owen, 34, 36, 39, 47, 94, 240n4
Low, Chris, 131
Lowen, C. B., 93
Lowie, Robert, 162
Lubbock, John, 158
Lyon, Stephen, 146–47

Machiavellian intelligence, 126, 137
male-bonding hypothesis, 26
male philopatry, 36
  primeval origins of, 36–41, 44–45, 175–80
Manger, Paul, 140–41
Marlowe, Frank, 194
Martin, Emily, 24–25
Marx, Karl, 158, 161
Maslin, Mark, 76, 78, 83
matriarchate, 162
matricentric family, 33, 44
  female feeding strategies and, 49–54
  mammalian origins of, 54–55, 85, 94–95
matrilineal organization
  as aberrant or unnatural, 172–73, 177, 192–93, 195, 203, 205, 211
  conditions favoring, 192–210
  cuckoldry and, 28, 172
  disappearance of, 207–210

matrilineal puzzle and, 202–7, 245n6
migration, warfare and, 194–99
primeval origins of, 84–96, 174–75, 177–82, 185–87, 232–33
regulatory mechanisms of, 223
resources, labor and, 200
Mayr, Ernst, 11
McBrearty, Sally, 129
McLennan, John, 158, 161, 180, 187
Mills, D., 3
Milton, K., 87
mirror neurons, 141
mitochondrial DNA, 186, 231, 244n11
Morgan, Elaine, 79–80
Morgan, Lewis H., 158, 161–63, 168–69, 180, 187
Morgan, T. J. H., 135
mosaic brain evolution, 139–43, 230–33
  energetics and, 152
  female philopatry and, 185–87, 231–32
  imprinted genes and, 183–87, 231
  linear scaling and, 139–41
  neurobehavioral specialization and, 124, 142, 146
multilevel selection, 5–6, 14, 21, 62–65
  epigenesis and, 125–26, 190
  eusociality and, 57–58
  group nepotism and, 189
Murdock, George P., 66, 121, 164–65, 196, 207, 209–10, 218–19, 220, 222

"new synthesis" model, 160–61
niche construction, 7–8, 193
  matriliny and, 200–2
  patriliny and, 211–14
Nicholas, George, 100–1, 145
Noonan, Katherine, 34

O'Connell, J. F., 74–75, 84, 87, 181
Opie, Kit, 88, 181

Orians, Gordon, 49
Oring, Lewis, 49
Otterbein, Charlotte, 194
Otterbein, Keith, 194

pair bonding
  hominin evolution and, 39, 45–46, 48, 121–22, 176–79
  nuclear family and, 33–34, 121–22
  sexual contract and, 34, 174
Pan-genesis theory, 175–82
  critique of, 177–182
Parker, Hilda, 166–67
Parker, Seymour, 166–67
partible paternity, 122
patrilineal organization
  conditions favoring, 210–18
  internal warfare and, 194, 212–13
  labor, productivity and, 214–18
  primeval origins of, 35–48, 170–72, 175–80, 228–29
  regulatory mechanisms of, 224
  resource scarcity and, 212, 214–15, 224
Pavard, Samuel, 174
Pawlowskil, Boguslaw, 93
Pennisi, Elizabeth, 186, 244n11
Peterson, Dale, 62
Pinker, Steven, 139, 143
Piperno, Dolores, 106
Poewe, Karla, 214
polygynandrous mating, 120–24
Potts, Richard, 76
Powell, Adam, 129
Power, Camilla, 88, 181

Rabinovich, Rivka, 135
Rahman, Qazi, 125
Reader, Simon, 138, 147
Reed, Kaye, 76
Renfrew, Colin, 128
Richards, Audrey, 202–3
Rilling, James, 142
Rodseth, Lars, 36, 38–39, 45–46, 159
Ross, Alexander, 200
Ryan, Christopher, 30, 121–22, 124, 240n4

Sacks, Karen, 170
Sahlins, Marshall, 34, 36, 212
Sanday, Peggy, 168
sapient paradox, 128–29
savanna hypothesis, 71, 75–78, 80, 229
scavenging, 74, 85, 99, 100
Schick, K. D., 134
Schillaci, Michael, 93
Schlegel, Alice, 203–4
Schneider, David, 199, 203–4, 206
selfish gene theory, 6, 26
　female exploitation and, 27–28
Semendeferi, K., 139–40
Service, Elman, 36, 158–59, 194, 211, 215
sexual contract, 34, 38, 228. *See also* pair bonding
sexual selection
　male domestication and, 55, 61, 85, 174–75
　male prowess and, 55
　sexual dimorphism and, 93
Shea, John, 105
Shennan, Stephan, 129, 174, 206
Small, Meredith, 43
Smith, A. R., 99
Smith, Eric A., 174
Snodgrass, J. Josh, 87
Snow, D. R., 197
social brain theory, 137–38, 160–61
social demography, 143–51
social DNA, 8, 14–15, 21
　epigenesis and, 68, 125, 233
social scaling, 146
sociobiology, 5, 26–27, 79
Soffer, Olga, 105
spatio-temporal roadmapping, 117–20
Spencer, Herbert, 158
spindle cell neurons, 141
Stanley, S. M., 85
Steward, Julian, 36, 158–59, 211, 215
Stewart, Kathlyn, 81, 89, 149
Stiner, Mary, 103–5, 111–12, 151
stone tool-making
　language and, 134–37

Paleolithic traditions and, 11, 73–74
　subsistence and, 73–74, 100
Symons, Donald, 29, 34

Tanner, Nancy, 54–55, 61, 85, 93, 136
tetradic kinship model, 190–91
Thomas, Mark, 129
Tiger, Lionel, 15, 36, 62
Tobias, Philip, 80
Toth, N., 134
Trauth, Martin, 76
Trivers, R. L., 5
Tylor, Edward B., 65

universal evolutionary stages, 157–61, 175

van den Berghe, Pierre, 28–29, 171–74
van Schaik, Carel, 93, 152
van Velzen, H.U.E. Theden, 194
van Wetering, W., 194
von Bertalanffy, Ludwig, 153–54
von Economo neurons. *See* spindle cell neurons
Vrba, Elizabeth, 76

Watson-Franke, Maria-Barbara, 199, 204–5, 245n6
Wengrow, D., 129
Westermarck, E., 162
Weston, Gavin, 146–47
Wheeler, Peter, 92
White, Leslie A., 163
Wilkinson, Lawrence, 184
Wilson, Edward O., 5, 7–8, 26, 34, 55, 57–58, 62, 71–72
Wilson, M., 29
Wittenberger, James, 49
Wolpoff, M. H., 11
Wrangham, Richard, 49–50, 52–53, 62, 83–84, 116–17
Wynn, Thomas, 134–35

Zhou, W., 146
Zinner, Dietmar, 36, 40

www.ingramcontent.com/pod-product-compliance
Lightning Source LLC
Chambersburg PA
CBHW070913030426
42336CB00014BA/2399